Permeability Measurement of Tight Rock

Chaolin Wang · Yu Zhao

Permeability Measurement of Tight Rock

Chaolin Wang
Guizhou University
Guiyang, China

Yu Zhao
Guizhou University
Guiyang, China

ISBN 978-981-92-0596-7 ISBN 978-981-92-0597-4 (eBook)
https://doi.org/10.1007/978-981-92-0597-4

This Springer imprint is published by the registered company Springer Nature Singapore Pte Ltd.
The registered company address is: 152 Beach Road, #21-01/04 Gateway East, Singapore 189721, Singapore

Preface

The characterization of unconventional gas reservoirs, particularly shale formations, represents one of the most dynamic and challenging frontiers in contemporary geoscience and reservoir engineering. Permeability, as the governing parameter of fluid flow, is critical to the evaluation, development, and sustainable management of such resources. Yet, accurately measuring this property in ultra-low permeability media remains fraught with methodological constraints, physical complexities, and interpretive ambiguities.

This monograph is born out of both necessity and innovation. While conventional steady-state methods prove impractical for tight rocks, transient techniques such as the pressure-pulse decay method have become the standard. However, their application is often hindered by theoretical simplifications, experimental limitations, and the pervasive influence of gas adsorption—a phenomenon intrinsic to organic-rich shales and coals. Moreover, the anisotropic nature of sedimentary rocks further complicates the quest for representative permeability values, as traditional measurement approaches struggle to capture directional flow behavior without introducing significant error or requiring multiple, potentially non-comparable samples.

In response to these challenges, this work systematically advances the theory, methodology, and application of transient permeability measurement. It moves beyond conventional analysis by introducing novel scaling techniques and integrated experimental approaches that allow for the simultaneous determination of permeability, porosity, adsorption capacity, and compressibility from a single test. Through rigorous numerical validation and laboratory demonstration, the methods presented herein not only enhance measurement accuracy but also significantly improve efficiency—offering practical solutions to long-standing problems in petrophysical characterization.

Structured as a progressive exploration, the monograph begins with a comprehensive review of existing permeability testing methods, followed by the development of new experimental frameworks and analytical tools. Each chapter builds upon the last, culminating in integrated methodologies for anisotropic characterization and adsorption-coupled flow analysis. It is my hope that this work will serve as both a reference and an inspiration for researchers, engineers, and students engaged in the science of fluid flow in tight porous media. During writing this book, we made abundant references to key publications in related fields and tried to tell the readers in the latest manner. Due to the limitation of our knowledge, there are some inevitable mistakes and defects in the book. Your suggestions would be deeply appreciated.

I extend my sincere gratitude to colleagues, collaborators, and institutions whose support made this research possible. Special thanks are due to the technical teams and laboratory staff whose diligence ensured the quality of the experimental work, and to the reviewers whose insights strengthened the manuscript.

The quest for better methods to understand subsurface flow is unending. It is my earnest hope that this contribution will advance that quest, enabling more reliable reservoir evaluation, more efficient resource recovery, and more effective geological storage of greenhouse gases in the years to come.

Guiyang, China
December 2025

Chaolin Wang
Yu Zhao

Acknowledgements This book is supported by the Research Fund for Talents of Guizhou University (Grant No. 201901), and Specialized Research Funds of Guizhou University (Grant No. 201903). We are grateful to the College of Civil Engineering, Guizhou University, which provide us excellent work and researching environment.

We profits from the stimulating discussions and the helpful and open atmosphere in our research group constituted by talented master and Ph.D. students. In particular, our gratitude goes to *Dr. Kunpeng Zhang* who have contributed a great deal to the final form of this book. We would also like to express our gratitude to organizations for permitting us to reproduce some of the figures.

Our final thank goes to the publisher Springer Nature Singapore Pte Ltd. We acknowledge the excellent support of *Wayne Hu* (Publishing Editor) while working on the book manuscript.

Competing Interests The authors have no competing interests to declare that are relevant to the content of this manuscript.

Contents

Chapter 1
Introduction

1.1 Background and Motivation

Accurate characterization of petrophysical properties-most critically permeability, porosity, adsorption capacity, and compressibility is fundamental to understanding fluid flow and storage in subsurface reservoirs. This understanding is essential for a wide range of energy and environmental applications, including the evaluation and development of conventional hydrocarbon resources, the assessment of geological carbon sequestration potential, and the efficient extraction of unconventional resources such as shale gas and coalbed methane.

Traditional laboratory methods for determining these properties often involve separate, time-consuming experiments on multiple core samples. For instance, permeability is typically measured using steady-state or transient techniques like the pressure-pulse decay (PPD) method, porosity is obtained from helium expansion or saturation techniques, and adsorption isotherms are measured using volumetric or gravimetric analyzers. Such a segmented approach not only increases experimental time and cost but also introduces uncertainties due to potential heterogeneity between different core plugs. Furthermore, in tight rocks with ultra-low permeability, traditional methods face significant challenges, including long equilibration times, sensitivity to temperature fluctuations, and the complicating effects of gas adsorption and pore structure deformation under pressure.

Consequently, there is a pressing need in both academia and industry for robust, efficient, and integrated experimental methodologies. An ideal method would be capable of simultaneously determining multiple key petrophysical parameters from a single test on a single core sample, thereby enhancing accuracy, saving time and resources, and providing more consistent data for reservoir modeling.

C. Wang and Y. Zhao, *Permeability Measurement of Tight Rock*,
https://doi.org/10.1007/978-981-92-0597-4_1

1.2 The Pressure-Pulse Decay Method: A Foundation and Its Limitations

Since its introduction by Brace et al. (1968), the pressure-pulse decay (PPD) method has become a cornerstone technique for measuring the permeability of low-permeability rocks. Its principle involves creating a small pressure differential between two chambers connected by a core sample and monitoring the transient pressure decay as gas flows through the sample towards equilibrium. The decay curve is then analyzed, typically using analytical solutions to Darcy's law, to estimate permeability.

Despite its widespread adoption, the conventional PPD method has several well-documented limitations:

(1) Single-parameter estimation: The standard analysis primarily focuses on permeability determination, often requiring independent experiments for porosity and compressibility, which are treated as known inputs.
(2) Neglect of adsorption effects: In gas-bearing rocks like coals and shales, the sorptive storage and transport of gases (e.g., methane, CO_2) can significantly influence the pressure transient. Ignoring adsorption can lead to substantial errors in permeability estimation.
(3) Ambiguity in data interpretation: The analysis often hinges on identifying specific flow regimes (e.g., early-time versus late-time behavior) on semi-log plots, which can be subjective and sensitive to experimental noise.
(4) Assumption of constant compressibility: The method typically assumes constant pore volume compressibility, which may not hold true, especially under large pressure changes where rock deformation is non-linear.
(5) Limited configuration: The standard two-chamber setup, while effective, may not be optimal for all scenarios, such as when sample availability is limited or when testing under specific boundary conditions.

These limitations highlight critical gaps in our ability to fully characterize complex tight rocks using a single, efficient experiment.

1.3 Objectives and Scope of This Monograph

This monograph is dedicated to advancing the theory and practice of transient flow experiments for the comprehensive characterization of tight porous media. The central objective is to develop and validate a suite of novel and modified experimental methods and data interpretation techniques that overcome the key limitations of the conventional PPD approach.

The specific aims are to:

(1) Develop new theoretical frameworks and scaled parameters to more accurately describe the pressure propagation process, particularly distinguishing and utilizing early- and late-time data.
(2) Propose integrated methodologies for the simultaneous determination of multiple parameters (e.g., permeability, porosity, adsorption capacity, compressibility) from a single test.
(3) Introduce modified experimental configurations (e.g., one-chamber systems) to enhance flexibility and application scope.
(4) Extend these principles to novel applications, such as determining permeability anisotropy from production data.
(5) Provide rigorous validation through numerical simulation and experimental case studies on real rock samples.

1.4 Outline of the Monograph

The structure of this monograph is designed to systematically address the objectives outlined above.

Chapter 2 provides a comprehensive review of permeability testing methods, with a focus on the evolution, analytical models, and challenges of transient pulse decay techniques, setting the context for the subsequent innovations.

Chapter 3 presents a new approach to simultaneously determine permeability, porosity, and adsorption capacity. It introduces a novel scaled pressure analysis to clarify transient behavior and a new data-processing method that decouples and quantifies these key properties while identifying the appropriate adsorption model.

Chapter 4 describes a modified pressure-pulse decay method using one chamber, offering a simplified experimental setup for the simultaneous measurement of adsorption and permeability.

Chapter 5 develops a large pressure pulse decay method to simultaneously measure permeability and compressibility of tight rocks. This chapter addresses rock deformation effects by employing large pressure pulses and a corresponding numerical model to extract pore volume compressibility alongside permeability.

Chapter 6 proposes a method for simultaneously determining axial and transverse permeability by canister degassing test. This chapter extends the transient analysis concept to a field-like scenario, using production data from a canister-degassing experiment to estimate permeability anisotropy.

Through this progressive exploration, this work aims to provide researchers and engineers with a refined toolkit for petrophysical evaluation, pushing the boundaries of what can be learned from a single, well-designed transient test.

BY NC ND

Chapter 2
Review of Permeability Testing Method

2.1 Introduction

Unconventional natural gas resources, encompassing shale gas, coalbed methane, and tight sandstone gas, have emerged as vital contributors to the global energy supply. The efficient development of these resources necessitates a thorough understanding of key reservoir properties, among which permeability is arguably the most critical parameter governing commercial productivity. While in-situ field testing provides the most representative measurements of reservoir permeability, it is inherently challenging to fully characterize gas permeability under dynamic field conditions. Permeability is not an intrinsic constant; it varies as a function of pore pressure and effective stress, which change throughout the production lifecycle. Furthermore, in organic-rich rocks like coal and shale, gas adsorption introduces additional complexity, significantly influencing both gas storage and flow behavior.

To construct reliable predictive models for gas production, it is essential to quantify how permeability responds to these controlling factors. Laboratory characterization serves as a powerful tool in this endeavor, allowing for the precise specification and control of experimental conditions. Through carefully designed experiments, it is possible to simulate various stages of gas production or injection, enabling a systematic investigation of permeability as a function of gas pressure, effective stress, adsorption, and temperature.

However, measuring the permeability of unconventional reservoir rocks presents significant technical challenges due to their characteristically low to ultra-low permeability, often defined as 0.001 millidarcy (mD) or less. Typical permeability ranges vary by rock type: coal exhibits permeabilities from microdarcy (μD) to hundreds of millidarcy, tight sandstones generally fall below 0.1 mD, and shales can range from nanodarcy (nD) to a few microdarcy, with some samples even demonstrating permeabilities below 1 nD, particularly in the direction perpendicular to bedding.

C. Wang and Y. Zhao, *Permeability Measurement of Tight Rock*,
https://doi.org/10.1007/978-981-92-0597-4_2

Given these extremely low values, permeability measurements can be exceedingly time-consuming, making the selection of an appropriate experimental methodology paramount. It is difficult to assign definitive applicability ranges to different techniques, as their effectiveness heavily depends on specific apparatus design and sample characteristics. Nonetheless, various methods-including steady-state, pulse decay, oscillating pressure, and the Gas Research Institute (GRI) method-are capable of measuring permeabilities in the nanodarcy range when properly configured.

This chapter aims to provide a comprehensive review of established and emerging permeability testing methods. Such an analysis is intended to aid in the design of improved experimental apparatus and procedures, thereby enhancing the characterization of ultra-low permeability rocks. A deeper understanding of the interplay between rock properties, gas interactions, and stress conditions will not only advance fundamental knowledge but also support the development of robust, universally applicable standards for low-permeability measurement.

2.2 The Steady State Method

The steady state method is the standard method to determine permeability of intact rock cores or plugs in the laboratory. Researchers have found it an accurate and reliable technique, with the advantage of a comparatively simple experimental set-up and a straightforward analytical solution. However, its application to very low permeability rocks is typically considered impractical, as it takes a long time to reach initial equilibrium during experiment preparation, and then more time to achieve steady state. The high degree of accuracy required of the instrumentation to measure very low flow rates associated with low permeability media also presents a practical challenge. With decreasing permeability, the time required for experiments increases, while the accuracy of the measurements decreases. Still, steady state experiments can be successfully performed for low and ultra-low permeability media, and provide reliable results.

A typical steady state experimental set-up is presented in Fig. 2.1. The sample, which is usually cylindrical, is held in a pressurised cell (often a Hassler or Hoek cell) and tightly wrapped in a rubber sleeve or a similarly flexible, impermeable material. Thin lead foil is used between the sample and the sleeve, if measurements are made with supercritical CO_2. This prevents any gas from flowing or diffusing out of the sample, and prevents any confining fluid from entering the sample. At either end of the sample is a stainless steel disc of the same diameter as the sample, which connects to (at least) one pump responsible for maintaining the sample's pore pressure or flow rate. The discs can be porous, or have perforations, engraved flow pathways or similar, to distribute the fluid across the sample's cross-sectional surface area. In the pressure cell, the sleeved sample is exposed to confining pressure. This is applied by pumping confining fluid, for example mineral oil or water, directly into the pressure cell. Confining pressure, upstreampressure (the end at which fluid is injected) and downstream pressure are measured using pressure transducers. The flow rate can be

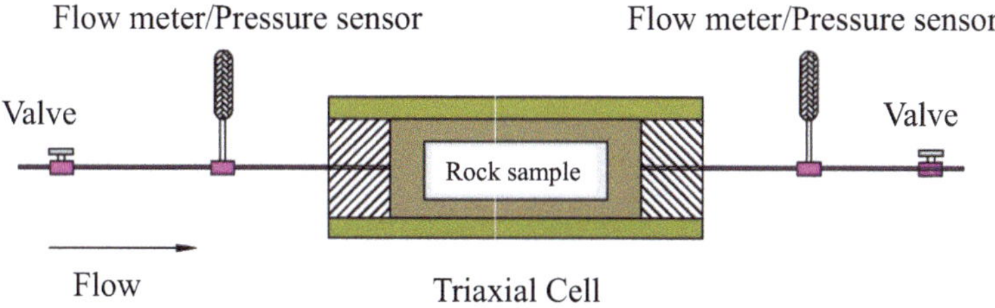

Fig. 2.1 The schematic sketch of the steady state method

measured directly via flow meters, or indirectly through volume changes over time in the upstream and downstream pumps. The rig must be temperature controlled, to ensure isothermal conditions that prevent undesired changes in pressure and flow rate during the experiments. This particularly applies to liquids for which small temperature changes can significantly distort pressures. Due to the very low flow rates associated with low permeability measurements, accurate instruments and controlled isothermal conditions are critical to obtaining reliable results. All measurements are recorded over time.

To carry out steady state experiments, the core is brought to the desired pore pressure and left to equilibrate. In the next step, either the pressure on the upstream side is increased, or the pressure on the downstream end is decreased. This creates a pressure differential across the length of the sample, which initiates flow. The experiment can be run by either maintaining a constant pressure differential until inflow equals outflow, or by imposing constant flow rate conditions and monitoring upstream and downstream pressure until the pressure differential becomes constant. McPhee and Arthur [1] compared the two methods using 160 mD sintered glass filter material, and determined that both yield comparable results. Similar findings were reported by Zhang et al. [2]. However, the constant pressure differential method is typically recommended for very low permeability media, because pressure measurements are more accurate than measurements of very low flow rates.

At steady state, Darcy's law can be applied to determine fluid flow through a porous medium. For incompressible fluids or near incompressible fluids, such as liquids, Eq. (2.1) can be applied directly and dp/dx becomes Δp/L, which is the pressure difference per unit of sample length.

$$Q = -\frac{kA}{\mu} \cdot \frac{dp}{dx} \tag{2.1}$$

For compressible fluids, such as gases, the fluid density varies as a result of the pressure gradient. Mass flow, however, is constant during steady state. Chen (1994) wrote Eq. (2.1) for incompressible fluids as [3]

$$m = -\frac{k \cdot A}{\mu} \cdot \frac{dp}{dx} \rho_f \tag{2.2}$$

in which m is the mass flow rate and ρ_f is the fluid density.

Through integration of Eq. (2.2) from inlet (upstream) to outlet (downstream) pressure and rearranging and substitution, Eq. (2.3) for the flow of compressible fluids is derived. The derivation is described in detail in Chen [3].

$$Q = -\frac{kA}{\mu L} \cdot \frac{p_d^2 - p_u^2}{2p_d} \tag{2.3}$$

where pd is the downstream pressure and pu is the upstream pressure.

In the derivation of Eq. (2.3), the gas is assumed to be ideal gas. Methane can be regarded as an ideal gas at low and medium pressures at reservoir temperatures. At elevated pressures, the real gas law is required to calculate permeability from experimental measurements. For this, a numerical solution may be used, because an analytical solution, such as Eq. (2.3), may not exist for real gas.

Jones and Meredith [4] and Boulin et al. [5] presented a push–pull configuration to determine the permeability of nanoporous materials (shales and clays, respectively, with permeabilities in the order of 10^{-22}–10^{-21} m2 (0.1–1 nd) using a constant pressure differential technique. The push–pull configuration consists of two piston pumps: one on the upstream end of the sample and the other on the downstream [5]. After the initial system equilibration, the pistons maintain the desired pressure differential by moving in a push (upstream piston) and pull (downstream piston) motion. The displacement of the pistons is plotted over time; knowing volume and time, the flow rate can then be determined. When the piston displacement curves become linear, the flow rate is constant, and permeability can be calculated applying Eq. (2.4).

$$k = -\frac{Q \cdot \mu \cdot L}{A \cdot \Delta p} \tag{2.4}$$

2.3 Unsteady State Methods

Because the steady-state method struggles to achieve a stable flow rate in tight rock samples, is time-consuming, and has low flow rate sensitivity, it is difficult to apply to dense formations such as shale and granite. As a result, transient methods have become one of the primary approaches for measuring permeability in tight rock samples. Currently, commonly used transient methods include the pressure pulse decay method, the GRI method, the oscillating pressure method, and the desorption method, among others.

2.3.1 *Pressure Pulse Decay Method*

The pressure pulse decay method was first proposed by Brace et al. in 1968. Figure 2.2 illustrates a schematic of the pressure pulse decay method. Initially, the experimental system is maintained at an initial equilibrium pressure. After the system reaches the initial equilibrium condition, a sudden pressure pulse is applied upstream (or the downstream pressure is abruptly reduced). Driven by the pressure gradient, fluid flows from the upstream reservoir through the rock sample to the downstream side. Permeability is calculated by monitoring the pressure difference between the upstream and downstream sides over time.

For the pressure pulse decay method, researchers have derived various theoretical solutions for permeability based on the assumption that fluid flow in the rock sample follows Darcy's law. These include solutions by Brace et al. [6], Hsieh et al. [7], Dicker and Smits [8], Jones [9], and Cui et al. [10], which can be summarized as follows:

(1) **Simplified Solution**

Brace assumed zero porosity for the rock sample and a constant pressure gradient in space, leading to a simplified solution for the pressure pulse decay method [6].

$$p_1 - p_f = \Delta p \frac{V_d}{V_u + V_d} e^{-\alpha t} \tag{2.5}$$

$$\alpha = \frac{kA}{c\mu L}\left(\frac{1}{V_u} + \frac{1}{V_d}\right) \tag{2.6}$$

where p_f represents the final equilibrium pressure, Δp denotes the initial pore pressure difference, V_u and V_d correspond to the volumes of the upstream and downstream containers respectively, μ is the fluid viscosity, and c represents the fluid compressibility.

Although Brace simplified the governing equations, the derived formula is straightforward and has consequently been widely adopted.

(2) **Early-Time Solution**

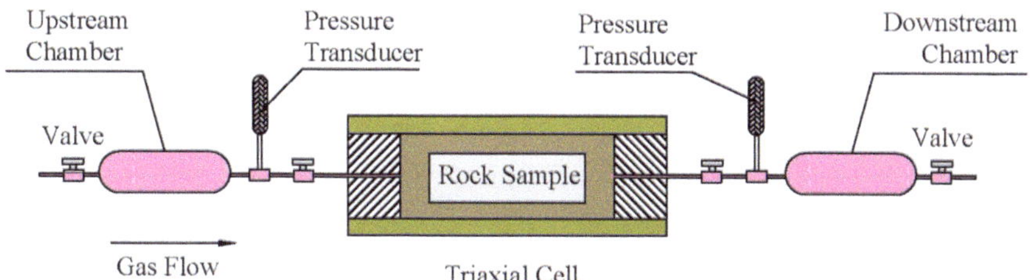

Fig. 2.2 The schematic sketch of the pressure pulse decay technique

After applying the pressure pulse, the pressure difference decreases rapidly. Theoretical solutions for this phase have been derived by Bourbie and Walls [11], Walder and Nur [12], Kamath et al. [13], He and Ling [14], among others. It should be noted that although the early-time solution can also be used to determine rock permeability, it is highly sensitive to rock anisotropy and is often affected by significant data noise during the early stage. Therefore, scholars generally recommend using late-time data for permeability calculations [9, 15–17].

(3) **Late-Time Solution**

Since the pressure difference-time curve exhibits an exponential decay relationship in the late-time phase, permeability can be calculated using the late-time pressure difference data.

(4) **Unified Solution**

To address the potential errors introduced by Brace's simplified assumptions, Hsieh et al. [7] derived an exact solution for the pressure pulse decay method. Dicker and Smits [8] further expressed Hsieh's solution as a unified function of dimensionless pressure and dimensionless time.

$$\Delta p_D = 2\sum_{m=1}^{\infty} \exp\left(-t_D\theta_m^2\right)\frac{a\left(b^2+\theta_m^2\right)-(-1)^m b\left[\left(a^2+\theta_m^2\right)\left(b^2+\theta_m^2\right)\right]^{0.5}}{\theta_m^4+\theta_m^2\left(a+a^2+b+b^2\right)+ab(a+b+ab)} \tag{2.7}$$

where θ_m is the solution to Eq. (2.6); a and b represent the ratios of pore volume to upstream container volume and downstream container volume, respectively; t_D denotes the dimensionless time, given by Eq. (2.9); and Δp_D represents the dimensionless pressure, given by Eq. (2.10).

$$\tan\theta = \frac{(a+b)\theta}{\theta^2 - ab} \tag{2.8}$$

$$t_D = \frac{k_a t}{c\mu\phi_a L^2} \tag{2.9}$$

$$\Delta p_D = \frac{p_1(t) - p_2(t)}{p_1(0) - p_2(0)} \tag{2.10}$$

Building on the research of Dicker, Jones [9] introduced a parameter "f" to reformulate the first term of Dicker's theoretical solution into a form consistent with Brace's solution, thereby enhancing the practical applicability of the unified solution.

Based on the transient pressure pulse method, researchers have extensively investigated the relationships between permeability and factors such as effective stress, moisture content, fracture characteristics, gas type, and temperature. For instance, Pini et al. [18] employed the pressure pulse method to study the influence of adsorption-induced swelling deformation on coal permeability under constant confining stress conditions. Pini first conducted permeability tests under helium gas conditions, establishing the relationship between permeability and effective stress. Specifically, as pore pressure increases, effective stress decreases, leading to an increase in permeability. The results showed that permeability increased by approximately 40 times when pore pressure rose from 1 to 7.4 MPa. Under nitrogen gas conditions, the trend of permeability variation with effective stress was consistent with that under helium conditions, though the magnitude of increase was smaller. In contrast, under carbon dioxide conditions, permeability initially decreased and then increased as effective stress decreased. Under identical experimental conditions, helium exhibited the highest permeability, followed by nitrogen, while carbon dioxide showed the lowest permeability.

Pan et al. [19] utilized the pressure pulse method to examine the relationship between permeability and effective stress. Their study revealed that under constant pore pressure conditions, gas permeability decreases exponentially with increasing effective stress. Chen et al. [20] applied the pressure pulse method to analyze the effects of the Biot coefficient in the effective stress formula and adsorption-induced matrix swelling deformation on rock permeability. To eliminate the influence of gas slippage, experiments were conducted under high pore pressure conditions. The findings indicated that the Biot coefficient in the effective stress formula is less than 1. Subsequently, Chen et al. [20] corrected the permeability results for methane and carbon dioxide based on the Biot coefficients determined at different pore pressures. The study demonstrated that assuming a Biot coefficient equal to 1 would exaggerate the impact of adsorption-induced swelling deformation on permeability, resulting in an overestimation of the permeability reduction caused by adsorption.

Wang et al. [21] applied the pressure pulse method to investigate the effects of fracture geometry (embedded fractures, axial through-going fractures, and circumferential through-going fractures) and moisture content on permeability under different test gases. The results indicated that for dry rock samples under constant confining pressure, the permeability of samples with embedded fractures and circumferential through-going fractures increased gradually with rising pore pressure under helium gas conditions. In contrast, under methane and carbon dioxide conditions, permeability initially decreased before increasing. For samples with axial through-going fractures, permeability consistently increased under all three gas conditions. Under water-saturated conditions, the permeability of all samples increased with pore pressure. Additionally, under identical experimental conditions, the permeability of dry rock samples was two orders of magnitude higher than that of water-saturated samples.

Zheng et al. [22] used the pressure pulse method to study the effects of gas type, pore pressure, effective stress, and temperature on rock permeability. Experiments were conducted using four gases: helium, nitrogen, methane, and carbon dioxide. The results showed that under identical conditions, the permeability of the four gases followed the order: helium > nitrogen > methane > carbon dioxide. When effective stress was held constant, the permeability of all four gases decreased with increasing pore pressure. Conversely, when pore pressure was held constant, the permeability of all gases decreased rapidly with increasing effective stress. Temperature-dependent permeability tests revealed that as temperature increased, the permeability of nitrogen and methane slightly decreased, while the permeability of carbon dioxide slightly increased.

Zhao et al. [23] employed the pressure pulse method to investigate the relationship between fracture morphology parameters and rock seepage characteristics. Additionally, Zhao et al. [17] studied the evolution of permeability under triaxial stress conditions and the relationship between fracture deformation and permeability.

Since its introduction, the pressure pulse decay method has undergone various refinements by researchers. Jones [9], aiming to reduce the initial system equilibration time, added two large-volume gas reservoirs to the traditional pressure pulse apparatus. This modification not only diminished the sensitivity of permeability measurements to rock anisotropy but also mitigated the impact of pore volume measurement errors on permeability results.

Metwally and Sondergeld [24], Heller et al. [25], and Civan et al. [26], among others, set the upstream storage capacity as infinite to maintain constant upstream pressure during pressure pulse tests. In this configuration, downstream pressure gradually increases, and permeability is calculated by monitoring the pressure difference over time. With this modified pressure pulse approach, the reliability of early-time solutions is significantly enhanced. Consequently, He and Ling [14] recommended using early-time solutions for permeability calculations with the modified method, substantially shortening experimental measurement durations.

Yang et al. [27] simplified the traditional pressure pulse method by proposing the use of only one chamber (the upstream chamber) for permeability testing and compared the results with those from conventional methods. Their findings indicated that the simplified approach not only streamlines the experimental setup and operation but also reduces the influence of pore volume measurement errors on permeability determination.

To address the excessively long testing times associated with tight rock samples such as shale, Hannon [28] introduced the bidirectional positive pulse method, which involves applying positive pressure pulses simultaneously to both upstream and downstream ends. He demonstrated that under identical experimental conditions, this method reduces testing time by a factor of 7.5 compared to conventional approaches.

Cao et al. [29] designed a pressure pulse apparatus with adjustable upstream and downstream chamber volumes to minimize permeability measurement errors caused by inaccuracies in pore volume estimation. This improved instrument allows for the adjustment of chamber volumes based on the measured pore volume of the core sample, thereby controlling the ratio of pore volume to chamber volumes.

Building on the pressure pulse decay method, Lasseux et al. [30] proposed a stepwise pressure pulse decay method. This technique employs two upstream chambers and one downstream chamber. The first step follows the conventional procedure, after which the second upstream chamber is used to apply pressure pulses incrementally, enabling stepwise pressure pulse testing. Experimental results demonstrated that the stepwise method provides more accurate measurements and allows for the simultaneous determination of absolute permeability and the slip factor of the rock sample.

The pressure pulse decay method has been widely used for measuring permeability in tight rock formations, yet it still faces two major challenges. Since early-time pressure difference data deviate from the exponential decay relationship, researchers generally recommend using late-time pressure difference data for permeability calculations. However, how to quantitatively define the boundary between early-time and late-time data remains an urgent issue in the practical application of the PPD method. Hsieh et al. [7] suggested that when the ratio of pore volume to chamber volume is less than 10, $t_D < 0.2$ could serve as a criterion for distinguishing between early-time and late-time pressure difference data. It should be noted, however, that t_D tself contains the unknown parameter permeability. Therefore, this criterion cannot be directly applied to permeability calculations.

When using adsorbable gases to measure the permeability of organic-rich rocks such as shale and coal, a portion of the free gas adsorbs onto the organic surfaces, providing additional gas storage capacity and thereby introducing errors into permeability measurements. To address this issue, Cui et al. [10] incorporated the Langmuir adsorption model into the traditional pressure pulse decay method, establishing a modified approach that accounts for adsorption effects. However, this method first requires determining the Langmuir adsorption parameters through separate adsorption tests.

Feng et al. [31, 32] attempted to eliminate the influence of storage compressibility and gas adsorption on permeability measurements by proposing a bidirectional pulse method (i.e., simultaneously increasing the upstream pressure by Δp while decreasing the downstream pressure by $-\Delta p$). Although their "positive–negative" bidirectional pulse method ensures that the final equilibrium pressure matches the initial equilibrium pressure exactly, it does not guarantee that the effects of storage compressibility and gas adsorption on the pressure difference data are neutralized during the test. Therefore, the applicability of the "positive–negative" bidirectional pulse method requires further investigation.

2.3.2 GRI Method (Gas Research Institute Technique)

The pressure decay method for rock cuttings, also known as the GRI method, was proposed by Luffel et al. [33] in 1992. This method involves testing rock cuttings under unstressed conditions and calculating the matrix permeability based on pressure decay data. The experimental setup for GRI permeability measurement mainly consists of a reference chamber and a sample chamber. Figure 2.3 illustrates the schematic diagram of the GRI method [10]. First, the rock sample is placed in the sample chamber, and gas pressure is applied to the reference chamber. After the pressure stabilizes, the valve connecting the reference chamber and the sample chamber is opened, allowing gas to expand from the reference chamber into the sample chamber. Due to the void volume in the system, the pressure in the reference chamber initially drops rapidly. Subsequently, as gas penetrates the rock pores, the pressure decline

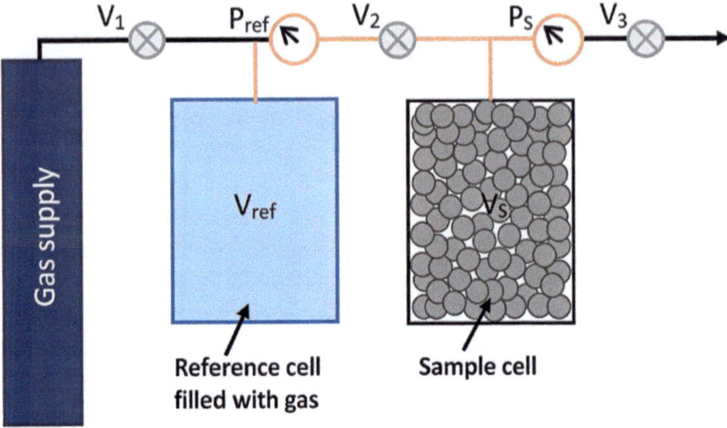

Fig. 2.3 The schematic sketch of GRI technique

slows until it stabilizes. By recording the pressure–time data in the sample chamber, the permeability of the rock sample can be calculated.

Compared to the transient method for core samples, the GRI method offers advantages such as shorter measurement time, lower cost, and the absence of microfractures in the cuttings, enabling accurate characterization of shale matrix permeability. However, the GRI method cannot simulate reservoir pressure, and the results are significantly influenced by the size and characteristics of the cuttings. Additionally, there is a lack of standardization in sample preparation, testing conditions, and calculation models. In 2009, Cui et al. [10] proposed a theoretical model for the GRI method that accounts for gas adsorption effects. In 2012, Tinni et al. [34] evaluated the GRI method by investigating the effects of particle size, sieving of crushed samples, pore pressure, gas type, and the initial state of the measurement apparatus on the experimental results. For tight reservoirs, Suarez-Rivera et al. [35] recommended that the particle size of cuttings should be between 0.84 and 1.68 mm to achieve optimal measurement time and repeatability. However, Cui et al. [36], Profice et al. [37], and Mokhtari [38], among others, argued that particle size significantly influences the experimental results.

2.3.3 Oscillating Pressure Method

The oscillating pressure method was proposed by Kranz et al. [39] in 1990 and can be used to measure rock permeability, diffusion coefficient, and storage capacity. The experimental setup of this method shares the same principle as the PPD method, with the key difference being that the oscillating pressure method applies a sinusoidal pressure variation upstream, rather than a constant pressure. Consequently, the pressure curve measured downstream also exhibits a waveform.

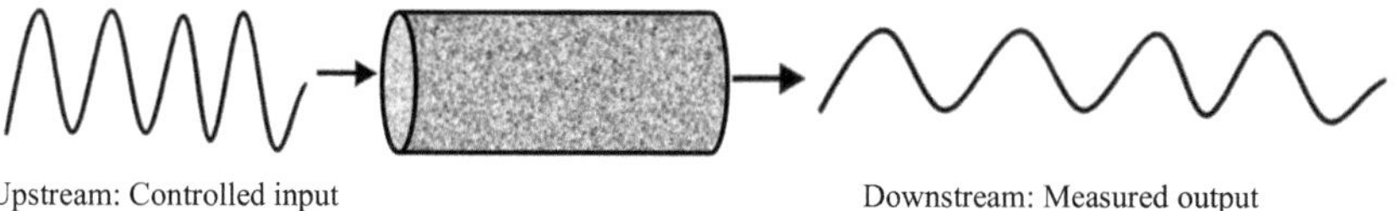

Fig. 2.4 The schematic sketch of the oscillating pressure method

First, a sinusoidal oscillating pressure wave with known amplitude and frequency is input upstream of the rock sample. After passing through the sample, an output sinusoidal wave of the same frequency is observed downstream, with its amplitude attenuated and phase delayed (as shown in Fig. 2.4). The amplitude attenuation and phase delay are related to the permeability of the rock sample. As the upstream oscillation frequency increases, the downstream amplitude decreases. Therefore, during experimental design, the upstream pressure should adopt optimal oscillation frequency and amplitude parameters. To minimize the influence of fluid compressibility on experimental results, Kranz et al. [39] suggested that the upstream amplitude should be less than 10% of the initial equilibrium pressure, while Wang and Knabe recommended it to be less than 5%. The selection of the optimal oscillation frequency depends on the hydraulic properties of the rock, fluid viscosity, and downstream container volume [39].

After its introduction, numerous scholars [40–45] have optimized and improved the oscillating pressure method. To enhance measurement accuracy, both Kranz et al. [39] and Fischer [46] recommended using rock samples with shorter dimensions and larger cross-sectional areas. Potters et al. [40] proposed using square wave signals for oscillating pulse tests, while Bernabé et al. [44] suggested reducing the downstream container volume.

2.3.4 Canister Degassing Test

In 1992, Luffel et al. [33] proposed the canister degassing test for measuring shale permeability. This method involves monitoring the gas desorption process from a core sample and determining the permeability based on pressure and production data. Compared to other transient methods, the desorption method measures cumulative flow versus time curves, and its mathematical processing is relatively complex. In 2009, Cui et al. [10] studied the mathematical processing of the canister degassing test for shale permeability measurement and provided permeability calculation methods corresponding to different sample sizes. Ettehadtavakkol and Jamali [47] proposed using the canister degassing test to measure shale matrix permeability and adsorption parameters, and conducted a comparative analysis of the effects of linear and nonlinear adsorption on the test results. Alfi et al. [48] employed the canister degassing test to measure shale permeability and validated the results by comparing them with the methods proposed by Javadpour [49] and Darabi et al.

[50]. Their study indicated that the permeability measurements obtained from the canister degassing test were more accurate.

2.3.5 Anisotropic Permeability Measurement

Shale, as a typical sedimentary rock, exhibits pronounced anisotropy, with permeability varying significantly depending on the direction relative to bedding planes. Zhang and Wang [51] defined shale permeability anisotropy as the ratio of permeability parallel to bedding to permeability perpendicular to bedding. Through experimental studies, they reported anisotropy values ranging from 1.2 to 1864.4 for the Longmaxi-Wufeng shale formations. Key factors contributing to anisotropic permeability in shale include mineral composition, grain size, pore volume, effective stress conditions, and natural micro-fractures [52–54]. Mokhtari and Tutuncu [38] conducted permeability tests on core samples with different bedding orientations using transient methods and found that bedding direction significantly influences shale permeability anisotropy.

Currently, research on methods for measuring anisotropic permeability in rocks remains limited. Some scholars, both domestically and internationally, have adopted approaches similar to conventional core testing, followed by mathematical processing to determine anisotropic permeability. For example, McCarney et al. [55] applied nuclear magnetic resonance (NMR) to measure rock permeability anisotropy; Luo et al. [56] proposed a method using TDS technology to determine permeability anisotropy; other techniques include acoustic logging for formation permeability anisotropy and dual-probe formation tester methods. However, most of these measurement techniques are tailored for rocks with relatively high permeability.

To obtain permeability values in different directions for shale reservoirs, researchers often measure permeability separately on samples with different bedding orientations. However, permeability results derived from different core samples may introduce significant errors. Studies by Kim and Lee [57], Zeynaly–Andabily and Rahman [58], and Soeder [59] indicate that permeability measurements from different samples at the same burial depth typically exhibit errors exceeding 50%. While methods for measuring axial permeability of core samples are relatively well-established, techniques for determining transverse permeability are less developed.

To address this gap, Yang et al. [15] developed a method for measuring transverse permeability based on transient principles. Their experimental setup applies pressure to the upper and lower end faces of the sample to ensure lateral gas flow, and transverse permeability is calculated by monitoring pressure changes in the sample chamber over time. Although this approach allows for separate measurement of transverse and longitudinal permeability on the same sample, it increases experimental workload and extends testing duration.

2.4 Key Existing Issues

(1) The transient method is the preferred approach for measuring shale permeability. However, the pressure difference decay curve in the pressure pulse method exhibits distinct early- and late-stage characteristics. Currently, late-stage data are predominantly used for permeability calculations, yet there is a lack of standardized methods and criteria for distinguishing between early and late stages. Furthermore, when using adsorbable gases such as methane and carbon dioxide for permeability testing, the influence of gas adsorption must be considered. At present, there is no effective method to account for or eliminate the impact of adsorption on permeability measurements.

(2) Influenced by geological structures and other factors, shale exhibits significant anisotropy. The permeability of shale can vary by up to two orders of magnitude depending on the bedding direction at the same burial depth. Currently, the common method for measuring permeability in different directions involves preparing cores from the same rock block in various orientations. However, sample preparation may cause varying degrees of damage, leading to potentially significant errors in permeability measurements across different bedding directions. How to accurately measure the anisotropic permeability of the same rock sample remains an area requiring further in-depth study.

(3) Existing shale permeability models typically focus on the effects of effective stress and adsorption/desorption-induced deformation on permeability, with limited consideration given to the influence of slippage effects. Additionally, experimental conclusions regarding the impact of temperature on permeability vary widely among scholars, and the underlying mechanisms of temperature effects on shale permeability remain unclear. Moreover, there is a scarcity of research on shale permeability models under temperature variations.

References

1. McPhee CA, Arthur KG (1991) Klinkenberg permeability measurements: Problems and practical solutions. In: Worthington PF, Longeron D (eds) Advances in core evaluation II reservoir appraisal. Proceedings of the 2nd society of core analysts European core analysis symposium. Gordon and Breach Science Publishers, Philadelphia, pp 371–391
2. Zhang M, Takahashi M, Morin RH, Endo H, Esaki T (2002) Determining the hydraulic properties of saturated, low-permeability geological materials in the laboratory: advances in theory and practice. In: Sara MN, Everett LG (eds) Evaluation and remediation of low permeability and dual porosity environments. ASTM International (ASTM STP 1415), pp 83–98
3. Chen G (1994) Gas slippage and matrix shrinkage effects on permeability of coal. The University of Arizona, USA. PhD thesis
4. Jones C, Meredith P (1998) An experimental study of elastic propagation anisotropy and permeability anisotropy in an illitic shale. Presented at the SPE/ISRM Eurock, Trondheim, Norway, 8–10 July 1998

5. Boulin PF, Bretonnier P, Gland N, Lombard JM (2012) Contribution of the steady state method to water permeability measurement in very low permeability porous media. Oil Gas Sci Technol 67(3):387e401
6. Brace WF, Walsh JB, Frangos WT (1968) Permeability of granite under high pressure. J Geophys Res 73(6):2225–2236
7. Hsieh P, Tracy J, Neuzil C, Bredehoeft J, Silliman S (1981) A transient laboratory method for determining the hydraulic properties of 'tight' rocks-I theory. Int J Rock Mech Min Sci Geomech Abs 18:245–252
8. Dicker A, Smits R (1988) A practical approach for determining permeability from laboratory pressure-pulse decay measurements. In: International meeting on petroleum engineering. Society of Petroleum Engineers
9. Jones SC (1997) A technique for faster pulse decay permeability measurements in tight rocks. SPE Form Eval 12:19–25
10. Cui X, Bustin AMM, Bustin RM (2009) Measurements of gas permeability and diffusivity of tight reservoir rocks: different approaches and their applications. Geofluids 9:208–223
11. Bourbie T, Walls J (1982) Pulse decay permeability: analytical solution and experimental test. SPE J 22:719–722
12. Walder J, Nur A (1986) Permeability measurement by the pulse-decay method: effects of poroelastic phenomena and non-linear pore pressure diffusion. Int J Rock Mech Min Sci Geomech Abs 23(3):225–232
13. Kamath J, Boyer RE, Nakagawa FM (1992) Characterization of core-scale heterogeneities using laboratory pressure transients. SPE Form Eval 7:219–227
14. He J, Ling K (2016) Measuring permeabilities of Middle-Bakken samples using three different methods. J Nat Gas Sci Eng 31:28–38
15. Yang Z, Dong M, Zhang S et al (2016) A method for determining transverse permeability of tight reservoir cores by radial pressure pulse decay measurement. J Geophys Res: Solid Earth 121:7054–7070
16. Feng R (2017) An optimized transient technique and flow modeling for laboratory permeability measurements of unconventional gas reservoirs with tight structure. J Nat Gas Sci Eng 46:603–614
17. Zhao Y, Wang CL, Zhao YL et al (2019) Experimental characterization and dependence of rock fracture permeability on 3D stresses. Arab J Geosci 12(2):41
18. Pini R, Ottiger S, Burlini L et al (2009) Role of adsorption and swelling on the dynamics of gas injection in coal. J Geophys Res: Solid Earth 114(B4):B04203
19. Pan ZJ, Connell L, Camilleri M (2010) Laboratory characterisation of coal reservoir permeability for primary and enhanced coalbed methane recovery. Int J Coal Geol 82(3):252–261
20. Chen ZW, Pan ZJ, Liu JS et al (2011) Effect of the effective stress coefficient and sorption-induced strain on the evolution of coal permeability: experimental observations. Int J Greenhouse Gas Control 5(5):1284–1293
21. Wang SG, Elsworth D, Liu JS (2011) Permeability evolution in fractured coal: the roles of fracture geometry and water-content. Int J Coal Geol 87(1):13–25
22. Zheng GQ, Pan ZJ, Chen ZW et al (2012) Laboratory study of gas permeability and cleat compressibility for CBM/ECBM in Chinese coals. Energy Explor Exploit 30(3):451–476
23. Yanlin Z, Lianyang Z, Weijun W et al (2017) Transient pulse test and morphological analysis of single rock fractures. Int J Rock Mech Min Sci 91:139–154
24. Metwally YM, Sondergeld CH (2011) Measuring low permeabilities of gas-sands and shales using a pressure transmission technique. Int J Rock Mech Min Sci 48(7):1135–1144
25. Heller R, Vermylen J, Zoback M (2014) Experimental investigation of matrix permeability of gas shales. AAPG Bull 98(5):975–995
26. Civan F, Rai CS, Sondergeld CH (2012) Determining shale permeability to gas by simultaneous analysis of various pressure tests. SPE J:717–726
27. Yang Z, Sang Q, Dong M et al (2015) A modified pressure-pulse decay method for determining permeabilities of tight reservoir cores. J Nat Gas Sci Eng 27:236–246

28. Hannon MJ (2016) Alternative approaches for transient-flow laboratory-scale permeametry. Transp Porous Media 114(3):719–746
29. Cao C, Li T, Shi J et al (2016) A new approach for measuring the permeability of shale featuring adsorption and ultra-low permeability. J Nat Gas Sci Eng 30:548–556
30. Lasseux D, Jannot Y, Profice S, Mallet M, Hamon G (2012) The 'Step Decay': a new transient method for the simultaneous determination of intrinsic permeability, Klinkenberg coefficient and porosity on very tight rocks. Presented at the International symposium of the society of core analysts, Aberdeen, Scotland, 27–30 Aug 2012
31. Feng R, Liu J, Harpalani S (2017) Optimized pressure pulse-decay method for laboratory estimation of gas permeability of sorptive reservoirs: part 1-background and numerical analysis. Fuel 191:555–564
32. Feng R, Harpalani S, Liu J (2017) Optimized pressure pulse-decay method for laboratory estimation of gas permeability of sorptive reservoirs: part 2-experimental study. Fuel 191:565–573
33. Luffel DL, Guidry FK (1992) New core analysis methods for measuring reservoir rock properties of Devonian shale. J Pet Technol 44(11):1184–1190
34. Tinni A, Fathi E, Agarwal R et al (2012) Shale permeability measurements on plugs and crushed samples. In: SPE Canadian unconventional resources conference. Society of Petroleum Engineers
35. Suarez-Rivera R, Chertov M, Willberg D, Green S, Keller J (2012) Understanding permeability measurements in tight shales promotes enhanced determination of reservoir quality. Presented at the SPE Canadian unconventional resources conference, Calgary, Alberta, Canada, 30 Oct–1 Nov 2012
36. Cui A, Brezovski R (2013) Laboratory permeability and diffusivity measurements of unconventional reservoirs: useless or full of information? A Montney example from the Western Canadian Sedimentary Basin. Presented at the SPE Asia pacific unconventional resources conference and exhibition, Brisbane, Australia, 11–13 Nov 2013
37. Profice S, Lasseux D, Jannot Y, Jebara N, Hamon G (2011) Permeability, porosity and Klinkenberg coefficient determination on crushed porous media. Presented at the International symposium of core analysts, Austin, Texas, USA, 18–21 Sept 2011
38. Mokhtari M, Tutuncu AN (2015) Characterization of anisotropy in the permeability of organic-rich shales. J Petrol Sci Eng 133:496–506
39. Kranz RL, Saltzman JS, Blacic JD (1990) Hydraulic diffusivity measurements on laboratory rock samples using an oscillating pore pressure method. Int J Rock Mech Min Sci Geomech Abs 27(5):345–352
40. Potters M, Mansoori M, Bombois X, Jansen JD, Van den Hof PMJ (2016) Optimal input experiment design and parameter estimation in core-scale pressure oscillation experiments. J Hydrol 534:534–552
41. Wang Y, Knabe RJ (2011) Permeability characterization on tight gas samples using pore pressure oscillation method. Petrophysics 52(6):437–443
42. Crawford BR, Faulkner DR, Rutter EH (2008) Strength, porosity, and permeability development during hydrostatic and shear loading of synthetic quartzclay fault gouge. J Geophys Res 113:B03207
43. Faulkner DR, Rutter EH (2000) Comparisons of water and argon permeability in natural clay-bearing fault gouges under high pressure at 20 °C. J Geophys Res 105:16415–16426
44. Bernabe Y, Mok U, Evans B (2006) A note on the oscillating flow method for measuring rock permeability. Int J Rock Mech Min Sci 43:311–316
45. Song I, Renner J (2007) Analysis of oscillatory fluid flow through rock samples. Geophys J Int 170:195–204
46. Fischer GJ (1992) Chapter 8—the determination of permeability and storage capacity: pore pressure oscillation method. In: Fault mechanics and transport properties of rocks. Academic Press Ltd
47. Ettehadtavakkol A, Jamali A (2016) Measurement of shale matrix permeability and adsorption with canister desorption test. Transp Porous Media 114(1):149–167

48. Alfi M, Hosseini SA, Enriquez D, Zhang T (2019) A new technique for permeability calculation of core samples from unconventional gas reservoirs. Fuel 235:301–305
49. Javadpour F (2009) Nanopores and apparent permeability of gas flow in mudrocks (shales and siltstone). J Can Pet Technol 48(8):16–21
50. Darabi H, Ettehad A, Javadpour F, Sepehrnoori K (2012) Gas flow in ultra-tight shale strata. J Fluid Mech 710:641–658
51. Zhang W, Wang Q (2018) Permeability anisotropy and gas slippage of shales from the Sichuan Basin in South China. Int J Coal Geol 194:22–32
52. Chalmers GR, Ross DJ, Bustin RM (2012) Geological controls on matrix permeability of Devonian gas Shales in the Horn River and Liard basins, northeastern British Columbia, Canada. Int J Coal Geol 103:120–131
53. Pan Z, Ma Y, Connell LD, Down DI, Camilleri M (2015) Measuring anisotropic permeability using a cubic shale sample in a triaxial cell. J Nat Gas Sci Eng 26:336–344
54. Ma Y, Pan Z, Zhong N et al (2016) Experimental study of anisotropic gas permeability and its relationship with fracture structure of Longmaxi Shales, Sichuan Basin, China. Fuel 180:106–115
55. McCarney ER, Butler PN, Taylor-Offord S et al (2015) Core plug nuclear magnetic resonance (NMR) analysis as a method to estimate permeability anisotropy. In: SPWLA 56th annual logging symposium. Society of Petrophysicists and Well-Log Analysts, Beach, California, USA
56. Luo Y, Long Z, Ma Z (2008) Determination of planar permeability anisotropy in reservoirs based on TDS technology. J China Univ Pet 05:63–66
57. Jang H, Lee W, Kim J et al (2016) Novel apparatus to measure the low-permeability and porosity in tight gas reservoir. J Pet Sci Eng 142:1–12
58. Zeynalyandabily EM, Rahman SS (1995) Measurement of permeability of tight rocks. Meas Sci Technol 6(10):1519–1527
59. Soeder DJ (1986) Laboratory drying procedures and the permeability of tight sandstone core. SPE Form Eval 1(1):16–22

Chapter 3
Analysis of the Pressure-Pulse Propagation in Rock: A New Approach to Simultaneously Determine Permeability, Porosity and Adsorption Capacity

3.1 Introduction

Permeability estimation of reservoir rocks is critical for applications in unconventional oil and gas recovery [1–3], geological storage facilities for CO_2 [4–6] and seals for nuclear waste repositories [7–9]. Currently, the steady-flow method and the transient-flow method are two major techniques for permeability measurement [10]. The steady-flow method consists of imposing a constant pressure gradient to a sample and measures the flow rates. However, the steady-flow technology is not adequate for tight rocks with low permeability, such as shale and higher rank coals, because the flow rates across the tight rock samples are often too small to be measured, and the tests are fairly time-consuming and inaccurate. Therefore, the transient flow method based on the pressure pulse decay signal which is easier to measure, becomes the preferred option.

The pressure pulse decay technique was initially proposed by Brace et al. [11] for the determination of permeability in tight rocks. This technique is based on the analysis of the differential pressure between the upstream and downstream (Fig. 3.1). In a pulse decay test, the whole test system is kept at a uniform pore pressure for a period to get an equilibrium condition at first. Then the pressure of upstream (or downstream) is increased (or decreased) to a specified level to create an initial pressure difference. The pressure responses at both ends are recorded after the valve is opened. The test ends when the upstream–downstream pressure difference becomes very small. The permeability can then be estimated from the pressure pulse decay curve (the upstream–downstream pressure difference with time) using mathematical solutions of the related flow problem. The standard exponential decay solution (often called as the Brace's solution) was first developed by Brace et al. [11] by assuming that the pressure gradient is constant along the length of the sample although it is a function of time, and that the pore volume of the sample can be neglected. Hsieh et al. [12] obtained a general solution of the 1-D flow problem without the simplifications used by Brace. Dicker and Smits [13] expressed the general solution in

C. Wang and Y. Zhao, *Permeability Measurement of Tight Rock*,
https://doi.org/10.1007/978-981-92-0597-4_3

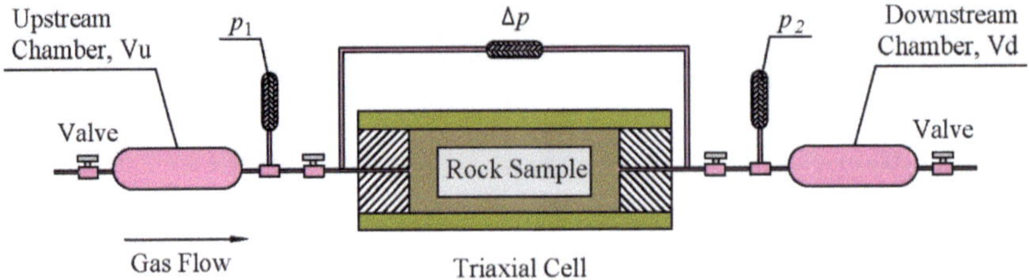

Fig. 3.1 The schematic sketch of the pressure pulse decay technique. The sample is first in equilibrium and then open the valve between the upstream chamber and the sample to let fluid flowing from the Upstream chamber to the Downstream chamber through the sample driven by the Δp

terms of dimensionless time and pressure difference for easier analysis. Jones [14] introduced a factor "f" to make Dicker and Smits's later-time solution in a similar form of Brace's solution. Recently, some optimized pressure decay methods were developed. Metwally and Sondergeld [15] implemented a new technique to simultaneously measure the permeability and storage parameters of tight rock samples by creating an infinite storage capacity for the upstream chamber. Yang et al. [16] suggested that only one chamber should be used in pulse pressure tests to simplify the experiment configurations. Hannon [17] provided a bidirectional model of pressure-pulse-decay permeametry, which reaches equilibrium much faster than the standard pressure-pulse decay.

The estimation of permeability from the measured pressure pulse decay data is complicated by the fact that the decay curve will often deviate from the single-exponential behavior in the early-time period (Fig. 3.2). Considerable error may result if early-time data are included for permeability calculation [18] because most of methods discussed above [14, 16, 18, 19] require that only the late-time measurements are used for permeability calculation and the early time data, although the early time measurements may contain important information about the rock samples [20–22], should be discarded. However, the time point differentiating the early-time and late-time behavior has not been clearly defined and was often somewhat arbitrary in literature. Although Hsieh et al. [12] found that for cases with the ratio of the compressive storage within the sample to the compressive storage in the upstream reservoir less than 10, the early time solution is applicable for the dimensionless time (defined in Eq. (3.14)) less than 0.2 while the later time solution is applicable for the dimensionless time larger than 0.2, this criterion cannot be used in practice because the dimensionless time itself contains the permeability which is unknown when we need to decide the subset of later time data.

The estimation of permeability from the measured pressure pulse decay data is further complicated due to possible existence of gas adsorption, a unique feature of kerogen-bearing rocks. In those rocks, free gas will lose because kerogen has a very high internal surface area for gas adsorption. In other words, the adsorption provides additional gas storage of the sample. Therefore, the estimated permeability from the pressure pulse decay method may be incorrect if adsorption effects are not considered

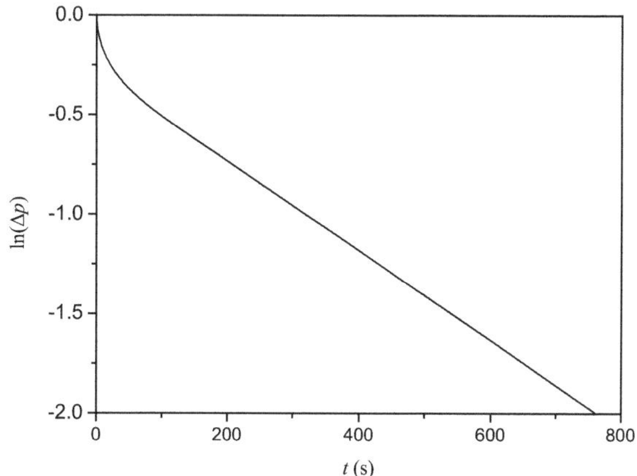

Fig. 3.2 A typical pressure pulse decay curve during pressure pulse decay experiment (Data come from Test 2 at confining pressure of 5 MPa in Sect. 3.3). The later- time upstream–downstream pressure difference shows exponential decay with time (i.e., straight line in semi-log plotting)

for those rocks with adsorbable gas. Cui et al. [23] proposed an approach by incorporating the Langmuir model into the Dicker and Smits' solution to consider adsorption effects. It becomes apparent that additional effort is required due to the increased number of input Langmuir parameters. Feng et al. [24] tried to eliminate the effect of compressibility storage and sorption on permeability measurement by introducing a bi-directional pressure pulse test (i.e. to create same magnitude but opposite direction of pressure pulses in each chamber concurrently). However, although the bi-direction pulse method could eliminate the net changes in compressive storage and sorption by equaling the initial pressure and the final pressure, the effects of spatially distributed adsorption/desorption on the upstream and the downstream pressure before ending of the pressure decay may not be properly cancelled each other so that the method was found not to be effective in removing the effects of sorption on the permeability measurement.

In this section, we presented a thorough analysis of the pressure pulse propagation process to reveal the mechanism behind the early time and later time behaviors of pressure decay curve and the reasons why the pressure decay curves are controlled by the governing equation using the apparent permeability and the apparent porosity if adsorption and gas slippage are existing. Inspired by the findings from these analyses, we proposed a new scaling method that can make the decay curves of all cases into a single 1:1 straight line for later time, which facilitate a new approach for inferring permeability and other related parameters from the measured pressure decay curve. The new approach can automatically select the later time data from the raw measurements and calculate the permeability and its standard deviation as well as apparent porosity and adsorption capacity. The new approach could not only remove the effects of the adsorption on the determination of the permeability but also identify proper adsorption model and parameters with multiple pressure pulse tests. We then performed a series of numerical simulations of pressure decay experiments with various rock properties under typical experiment settings to evaluate the accuracy and the robustness of the proposed approach. Finally, the new approach (and the associated code, documented in Appendix) is successfully applied to processing the real data obtained from several pressure decay experiments involving two types of rocks and two types of gases under different confining pressure and pore pressure.

3.2 The Principles of Pressure Pulse Decay Method: Apparent Porosity and Permeability

The pressure pulse decay method involves a closed system that consists of two chambers, under different initial pressures, which are connected through the rock core (Fig. 3.1). The initial pressure pulse will propagate through the rock core from the upstream chamber to the downstream chamber. The fundamental process can be seen as one-dimensional gas flow through the porous core sample, which can be described by the following governing equation:

$$\frac{\partial M}{\partial t} = \frac{\partial}{\partial x} \cdot \left(\frac{k_a \rho_g}{\mu} \frac{\partial p}{\partial x} \right) \tag{3.1}$$

where M is the total gas mass in the control volume, kg/m^3; ρ_g is the density of gas, kg/m^3; t is time, s; μ is gas viscosity, Pa s; p is the gas pressure, Pa; k_a is apparent permeability, m^2, which can be expressed as a general form including gas slippage effects [25];

$$k_a = k\left(1 + \frac{\lambda}{p_m}\right) \tag{3.2}$$

where k is true permeability, m^2; λ is Klinkenberg coefficient, Pa; p_m is a constant pressure, representing the mean gas pressure, Pa.

The gas mass in the control volume (M) consists of free-phase gas in pore space ($\phi \rho_g$) and adsorbed gas by rock matrix (the second term on the right of Eq. (3.3)), which can be described as follows:

$$M = \phi \rho_g + (1 - \phi) \rho_R \psi(p) \tag{3.3}$$

where ρ_R is the density of rock sample, kg/m^3; $\psi(p)$ is the mass of gas adsorbed per unit mass of rock, kg of gas/kg of matrix material; and $(1-\phi)\rho_R$ represents the mass of rock matrix in the control volume.

Substituting Eqs. (3.3) into (3.1) and assuming that spatial changes in ρ_g, k_a, and μ are negligible yields

$$\left[\phi \frac{d\rho_g}{dp} + \left(\rho_g - \rho_R \psi(p)\right) \frac{d\phi}{dp} + (1 - \phi) \rho_R \psi'(p)\right] \frac{\partial p}{\partial t} = k_a \frac{\rho_g}{\mu} \frac{\partial^2 p}{\partial x^2} \tag{3.4}$$

Considering that the changes in true porosity due to a small pressure pulse in test can be negligible (i.e. $\frac{d\phi}{dp} = 0$), the second term on the left side of Eq. (3.4) can be reasonably omitted. As a result, the governing Eq. (3.4) can be expressed as by introducing the concept of apparent porosity:

$$\phi_a \frac{\partial p}{\partial t} = \frac{k_a}{c\mu} \frac{\partial^2 p}{\partial x^2} \tag{3.5}$$

where the apparent porosity is a summation of true porosity, ϕ, and the equivalent porosity due to gas adsorption, ϕ_{ad}:

$$\phi_a = \phi + \phi_{ad} \tag{3.6}$$

And the equivalent porosity due to gas adsorption is a measure of the gas adsorption capacity of a sample in terms of "porosity" in Eq. (3.5). It can be defined as a function of the gradient of gas adsorption with respect to gas pressure:

$$\phi_{ad} = \frac{\rho_R(1-\phi)}{c\rho_g} \psi'(p) \tag{3.7}$$

where c is the gas compressibility $\left(c = \frac{1}{\rho_g} \frac{d\rho_g}{dp}\right)$.

Note that the equivalent porosity due to adsorption is a virtual porosity to account for the adsorption in the mass balance Eq. (3.7) that controls the gas transport in the rock sample. It is a mathematic equivalent variable representing the adsorption capacity in terms of porosity and there is no physical volume or density of adsorbed gas is involved in this concept. The equivalent porosity due to adsorption could be as large as many times of the true porosity, not necessary limited by the maximum (true) porosity of 1. Cui et al. [23] provided a number of examples of larger equivalent porosity although they called it as effective adsorption porosity. Unlike Cui et al. [23] who derived an equation to calculate the equivalent porosity based on Langmuir adsorption model, Eq. (3.7) here is a general relationship between the equivalent porosity and the adsorption. The concept of the equivalent porosity can be applied to all cases including Langmuir adsorption or even the case there is no closed form adsorption model exists. Gas ad-/desorption has been studied extensively and several analytic/semi-analytic models have been developed for predicting gas sorption behavior [26], such as Langmuir sorption, Freundlich sorption, BET (Brunauer, Emmett and Teller) sorption, and Dubinin-Astakhov (D-A) sorption. If the adsorption can be described by the Langmuir model, the mass of gas adsorbed ψ can be calculated as a function of pressure [27–29]:

$$\psi(p) = \frac{m_L p}{p + p_L} \tag{3.8a}$$

where m_L describes the maximum gas adsorption capacity, kg of gas/kg of matrix material; p_L is the pressure at which half of this capacity is reached, Pa; p is the pressure in the sample, Pa. In this case, the equivalent porosity can be calculated as:

$$\phi_{ad} = \frac{\rho_R(1-\phi)}{c\rho_g}\frac{m_L p_L}{(p+p_L)^2} \tag{3.8b}$$

The equations for calculating the equivalent porosity due to adsorption for other adsorption models are listed in Table 3.1.

Finally, the one-dimensional gas flow process in the pressure pulse decay method can be described by the governing Eq. (3.5) completed by the following initial and boundary conditions:

$$p(x,0) = p_2(0)\ \text{for}\ 0 < x < L \tag{3.9a}$$

$$p(0,t) = p_1(t)\ \text{for}\ t \geq 0 \tag{3.9b}$$

$$p(L,t) = p_2(t)\ \text{for}\ t \geq 0 \tag{3.9c}$$

$$\frac{dp_1}{dt} = \frac{k_a}{c\mu\phi_a L}\frac{V_S}{V_u}\frac{\partial p}{\partial x}|_{x=0}\ \text{for}\ t > 0 \tag{3.9d}$$

$$\frac{dp_2}{dt} = -\frac{k_a}{c\mu\phi_a L}\frac{V_S}{V_d}\frac{\partial p}{\partial x}|_{x=L}\ \text{for}\ t > 0 \tag{3.9e}$$

Table 3.1 The equations for calculating the equivalent porosity due to adsorption for adsorption models

Sorption model	Function	Equivalent porosity due to adsorption
Langmuir model	$\psi(p) = \frac{m_L p}{p_L + p}$	$\phi_{ad} = \frac{\rho_R(1-\phi)}{c\rho_g}\frac{m_L p_L}{(p+p_L)^2}$
Freundlich model	$\psi(p) = Kp^n$	$\phi_{ad} = \frac{\rho_R(1-\phi)}{c\rho_g}Kp^{n-1}$
BET model	$\frac{1}{\psi(p)(p_0/p-1)} = \frac{1}{V_m C} + \frac{C-1}{V_m C}\frac{p}{p_0}$	$\phi_{ad} = \frac{\rho_R(1-\phi)}{c\rho_g}\left\{\frac{Cp_0V_m p}{(p-p_0)^2(p_0-p+Cp)} - \frac{Cp_0V_m}{(p-p_0)(p_0-p+Cp)} + \frac{Cp_0V_m p(C-1)}{(p-p_0)(p_0-p+Cp)^2}\right\}$
D-A model	$\psi(p) = V_0 \exp\left[-D\left\{\ln\left(\frac{p_0}{p}\right)^2\right\}\right]$	$\phi_{ad} = \frac{\rho_R(1-\phi)}{c\rho_g}\frac{2DV_0}{p}\exp\left[-D\left\{\ln\left(\frac{p_0}{p}\right)^2\right\}\right]$

where x denotes the distance along the sample from the upstream end, m; L refers sample length, m; V_s, V_u, and V_d refer the volume of sample pore, upstream chamber and downstream chamber, m^3, respectively; p_1 and p_2 are the pressure of upstream chamber and downstream chamber, Pa.

Note that Eqs. (3.5) and (3.9a–e) have the same mathematical forms of the governing equations first presented by Brace et al. [11]. Therefore, the Brace's simple solution, as well as Hsieh's general solutions, can be readily presented as solutions of Eq. (3.5) by replacing porosity and permeability with apparent porosity and apparent permeability. In other words, in general, the permeability deduced from the pressure pulse decay method is an apparent permeability and the porosity used in calculation of permeability should be the apparent porosity. The conventional practice that uses porosity obtained from other methods may lead to significant errors in permeability, especially if the adsorption is significant.

With the apparent permeability, the Brace's solution can be written as:

$$p_1 - p_f = \Delta p \frac{V_d}{V_u + V_d} e^{-\alpha t} \tag{3.10}$$

in which

$$\alpha = \frac{k_a A}{c\mu L}\left(\frac{1}{V_u} + \frac{1}{V_d}\right) \tag{3.11}$$

where p_f is the final equilibrium pressure, Pa; Δp is the initial pressure difference, Pa; A is sample cross-sectional area, m^2; and k_a is the apparent permeability (instead of permeability in original Brace's solution), m^2. Brace's solution (Eqs. (3.10) and (3.11)) is the first solution and its simplicity helped the pressure pulse decay method to be widely used to measure the permeability of tight rock samples. However, the assumption of constant spatial pressure gradient in the rock sample $\left(\frac{\partial p^2(x,t)}{\partial x^2} = 0\right)$ may lead to significant permeability error in some cases which will be discussed later.

The more general Hsieh's solution [12] can be expressed in terms of dimensionless differential pressure as suggested by Dicker and Smits [13] with modified symbols:

$$\Delta p_D = \sum_{m=1}^{\infty} S_m = 2\sum_{m=1}^{\infty} \exp\left(-t_D\theta_m^2\right) \frac{a\left(b^2+\theta_m^2\right) - (-1)^m b\left[\left(a^2+\theta_m^2\right)\left(b^2+\theta_m^2\right)\right]^{0.5}}{\theta_m^4 + \theta_m^2\left(a + a^2 + b + b^2\right) + ab(a + b + ab)} \tag{3.12}$$

where S_m is the mth term, θ_m are the roots of Eq. (3.13); a and b are the ratios of apparent pore volume to upstream and downstream volume respectively; t_D is dimensionless time, defined in Eq. (3.14) and Δp_D is the dimensionless pressure difference, defined in Eq. (3.15):

$$\tan\theta = \frac{(a+b)\theta}{\theta^2 - ab} \tag{3.13}$$

$$t_D = \frac{k_a t}{c\mu\phi_a L^2} \tag{3.14}$$

$$\Delta p_D = \frac{p_1(t) - p_2(t)}{p_1(0) - p_2(0)} \tag{3.15}$$

Here the apparent porosity, ϕ_a, and the apparent permeability, k_a, are used instead of the porosity and permeability used in the original solutions.

Although the analytical solution Eq. (3.12) is an exact solution of the problem including both an early-time behavior and a late-time behavior, it is difficult to handle because of involving evaluation of infinite series. Therefore, the first term of Eq. (3.12) is often recommended for routine permeability calculation [14], because the late-time solution is dominated by the first term (i.e., $m = 1$) in Eq. (3.12) as shown in Fig. 3.3. The other terms ($m > 1$) only significantly affect the solution at early time and cause the solution deviating from the linear line described by the first term in a semi-log plotting (Fig. 3.3). Such effects of the higher order terms on the solution decrease with decreasing of the parameter a (the ratio of pore volume in the sample over the volume of the upstream chamber, $a = b$ here). However, the solution itself does not tell how to distinguish the early time and the later time measurement data.

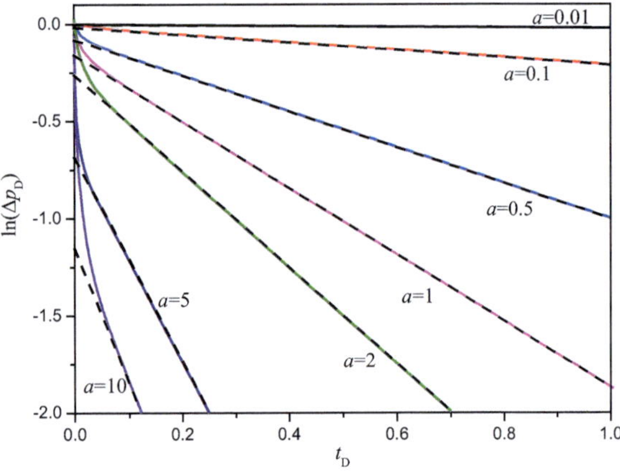

Fig. 3.3 A comparison of the general solution (Eq. (3.12), the solid curves) with the first term (the dashed curves) for different ratios of pore volume to chamber volume (assuming the volume of upstream chamber equals to that of the downstream chamber, i.e., $a = b$)

3.3 Numerical Analysis of Pressure Pulse Propagation Process: Early and Later Time Behaviors

As described in Fig. 3.1, the pressure pulse decay method is basically an experiment of pressure pulse propagation through the rock sample due to fluid flow from the upstream (higher pressure) chamber into the downstream (lower pressure) chamber. To understand the dynamics of such pressure pulse propagation through the various rock samples under different experimental parameters, we performed numerical simulations of 18 cases that cover variations in typical rock properties and the experiment configurations (e.g., *a* and *b*). Table 3.2 shows the parameters used in these numerical simulations. Taking Case 1 as the base case, a series of numerical simulations are performed at varied values of chamber volumes (Case 2 through Case 5 for upstream chamber volumes and Case 6 through Case 8 for downstream chamber volumes), rock porosity (Case 9 through Case 11), permeability (Case 12 and Case 13), adsorption parameters (Case 14 through Case 16) and slippage parameters (Case 17 and Case 18) (Table 3.2). The initial pressure is set to 1 MPa and the pressure pulse is set to 0.1 MPa for all numerical tests. Note that Langmuir sorption model is used in Case 14 to Case 16 to consider adsorption effects.

Table 3.2 Modeling parameters of numerical tests

Numerical tests	V_u (cm^3)	V_d (cm^3)	ϕ	a	b	k*10^{-18} (m^2)	m_L	p_L (MPa)	λ (MPa)
Case 1	98.2	98.2	0.05	0.1	0.1	1	/	/	/
Case 2	**982**	98.2	0.05	0.01	0.1	1	/	/	/
Case 3	**9.82**	98.2	0.05	1	0.1	1	/	/	/
Case 4	**4.91**	98.2	0.05	2	0.1	1	/	/	/
Case 5	**0.982**	98.2	0.05	10	0.1	1	/	/	/
Case 6	98.2	**982**	0.05	0.1	0.01	1	/	/	/
Case 7	98.2	**49.1**	0.05	0.1	0.2	1	/	/	/
Case 8	98.2	**9.82**	0.05	0.1	1	1	/	/	/
Case 9	98.2	98.2	**0.01**	0.02	0.02	1	/	/	/
Case 10	98.2	98.2	**0.1**	0.2	0.2	1	/	/	/
Case 11	98.2	98.2	**0.25**	0.5	0.5	1	/	/	/
Case 12	98.2	98.2	0.05	0.1	0.1	**0.01**	/	/	/
Case 13	98.2	98.2	0.05	0.1	0.1	**100**	/	/	/
Case 14	98.2	98.2	0.05	0.1	0.1	1	**0.001**	**5**	/
Case 15	98.2	98.2	0.05	0.1	0.1	1	**0.005**	**5**	/
Case 16	98.2	98.2	0.05	0.1	0.1	1	**0.01**	**5**	/
Case 17	98.2	98.2	0.05	0.1	0.1	1	/	/	**0.5**
Case 18	98.2	98.2	0.05	0.1	0.1	1	/	/	**1**

In our numerical simulations, the core sample was discretized uniformly into 100 grid cells and two chambers were represented as two special grid cells attached to the upstream and downstream ends, respectively. The chamber grid cells have exact volumes as specified in each case with unit porosity and large permeability. The sample dimensions are representative of actual core tested in the laboratory flow tests, that is, 50 mm in diameter and 100 mm in length. CH_4 is used for the numerical simulations.

These simulations were completed using TOUGH + REALGASBRINE (TOUGH+), a widely used numerical simulator for non-isothermal multiphase flow (an aqueous phase and a real gas mixture) in a gas-bearing medium, with a particular focus in ultra-tight systems [28]. Different from the analytical solutions discussed above, TOUGH + model does not assume constant gas density, viscosity, and compressibility. Instead, these properties are functions of pressure and temperature based on the real gas EOS model implemented in TOUGH+. Therefore, they will vary with space in the rock sample following the pressure as in reality although such variations are expected to be small for the given pressure difference in typical pressure pulse decay experiments. In addition, TOUGH+ adopts instant equilibrium adsorption/desorption (e.g., Langmuir model). Fewer simplifications make the numerical simulator to be more suitable than the analytical solutions to understand the dynamics and mechanism of the pressure pulse propagation taking placing in a pressure pulse decay experiment.

Figure 3.4 shows semi-log plotting of dimensionless pressure difference vs. dimensionless time for all cases. In general, the dimensionless pressure-difference decays quickly at early time and then approaches exponentially decay mode of dimensionless time (shown as straight lines in the plotting) after certain time point although the slope may vary greatly between cases.

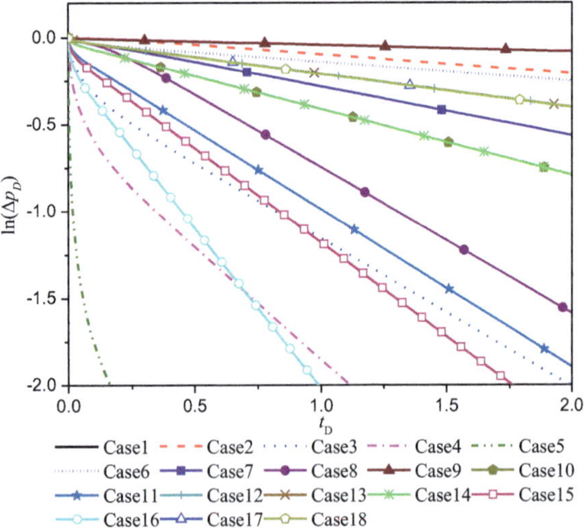

Fig. 3.4 Pressure responses in various cases (Case 1 is overlapped by the Case 12, Case 13, Case 17 and Case 18 so that the black for Case 1 cannot be seen)

To understand the mechanism behind such big changes in pressure decay pattern between early time and later time, we plot the contour of the normalized pressure change (defined as $\frac{p(x,t)-p(x,0)}{p(0,0)-p(L,0)}$) in temporal and spatial domain in Fig. 3.5 to see what happens inside the rock sample during the experiment. Each figure shows the evolution of pressure profile within the rock sample over time for each case. Breakthrough curve (the dashed line in Fig. 3.5) is defined as the normalized pressure reaches to 0.005 to describe the position of the front of the pressure pulse in the rock sample. In general, for the given pressure difference, velocity of the pulse propagation would be proportional to the permeability of the rock but inversely proportional to the storage terms (e.g., the pore volume of the rock, adsorption capability and the fluid compressibility). For example, with two orders increases in permeability among Cases 12, 1, and 13, the breakthrough time decreases from 65530 to 510 s to 7 s, respectively, in the same orders of the permeability. While the porosity increases from 0.01 to 0.05, 0.1 and finally to 0.25, the breakthrough time increases from 120 to 510 s, 1010 s and finally to 2010 s, almost in the same multiple growths of the porosity. The effects of adsorption and slippage on breakthrough time are similar to the results of porosity and permeability due to the changes of apparent porosity and apparent permeability. Another important factor is upstream chamber volumes. For example, the breakthrough time is about 510 s in Case 1, 900 s in Case 4, and 2370 s in Case 5 as the upstream chamber volume decreases from 98.2 to 4.91, to 0.982 cm^3, respectively. This is because a larger upstream chamber can hold more gas at the given pressure than a smaller one so that its pressure drop due to the same amount of gas loss to the rock sample tends to be smaller. As a result, a larger upstream chamber can keep a larger driving force for flow through the rock sample, which results in fast breakthrough of the pressure pulse. The downstream volume has little effects on the breakthrough time (Cases 6 through Case 8) because the downstream chamber has no effects on the pressure pulse propagation before the pressure pulse breakthrough the rock sample.

The physical process of the pressure pulse propagation is different before and after the breakthrough time. Before that time the dissipation of the pressure at the upstream chamber behaves as if the pressure pulse is propagating through an infinite medium, which is quite different from the situation after the pressure pulse reaches the downstream chamber so that the downstream boundary starts to affect the process. This difference in the process of the pressure pulse propagation results in different behaviors of the measured pressure response (usually expressed as pressure difference between the upstream and downstream chambers) at early time and later time. The turning point is the time the pressure pulse reaches the downstream end. After that breakthrough time, the flow becomes "quasi-steady", in which the pressure distributions within the rock quickly approach the constant gradient patterns [12].

To further show the evolution of pressure gradient over time, we plotted the difference in pressure gradient at two ends of the rock sample (Fig. 3.6). In principle, this difference shall be the maximum at time zero, gradually decreases with time and finally reaches zero because the system reaches the equilibrium state (e.g., Fig. 3.6a). As shown in Fig. 3.6b, although the pressure gradient difference decreases in quite a different rate among different cases, it approaches zero (<0.005) for all cases when

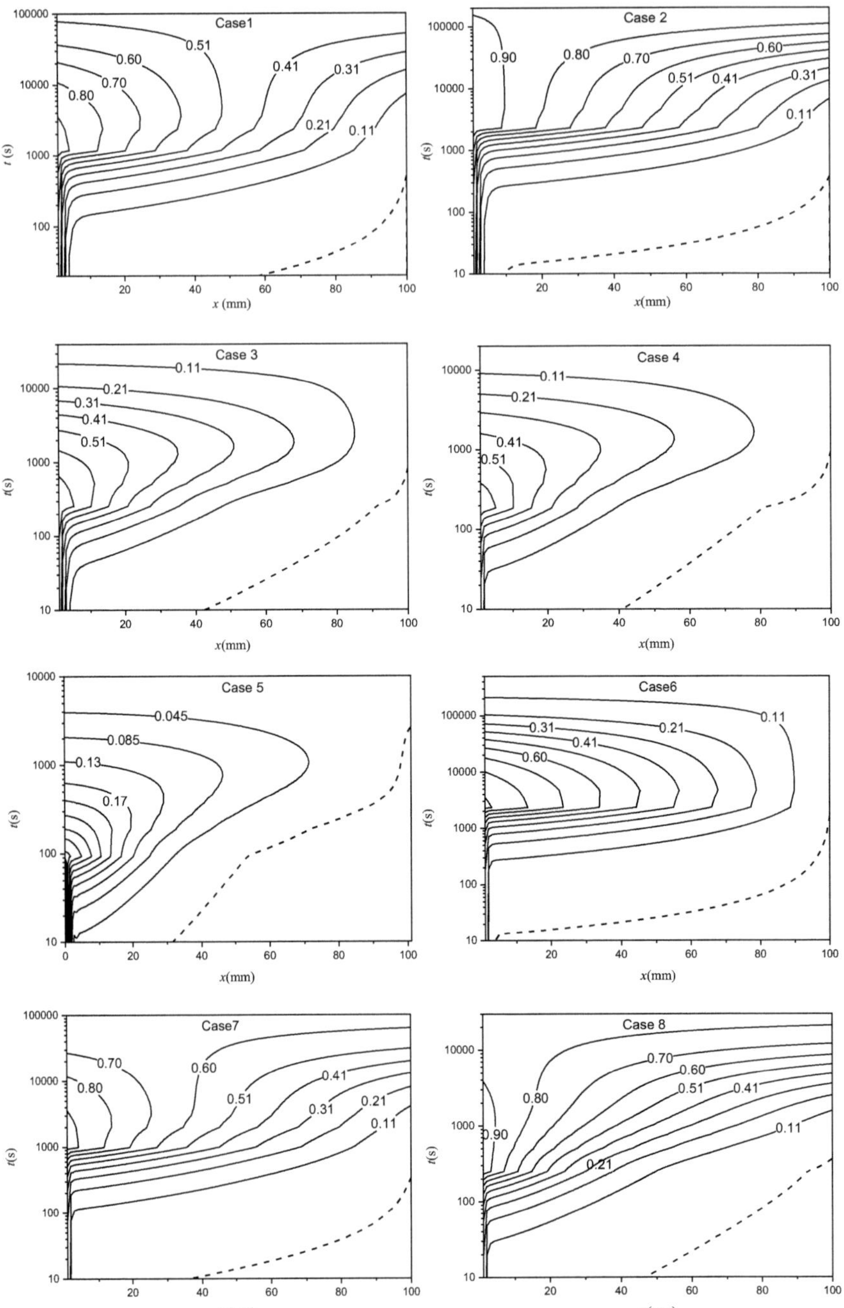

Fig. 3.5 Contours of the normalized pressure change in space and dimensionless time domain for Cases 1 to 18. The dashed lines describe the breakthrough curve

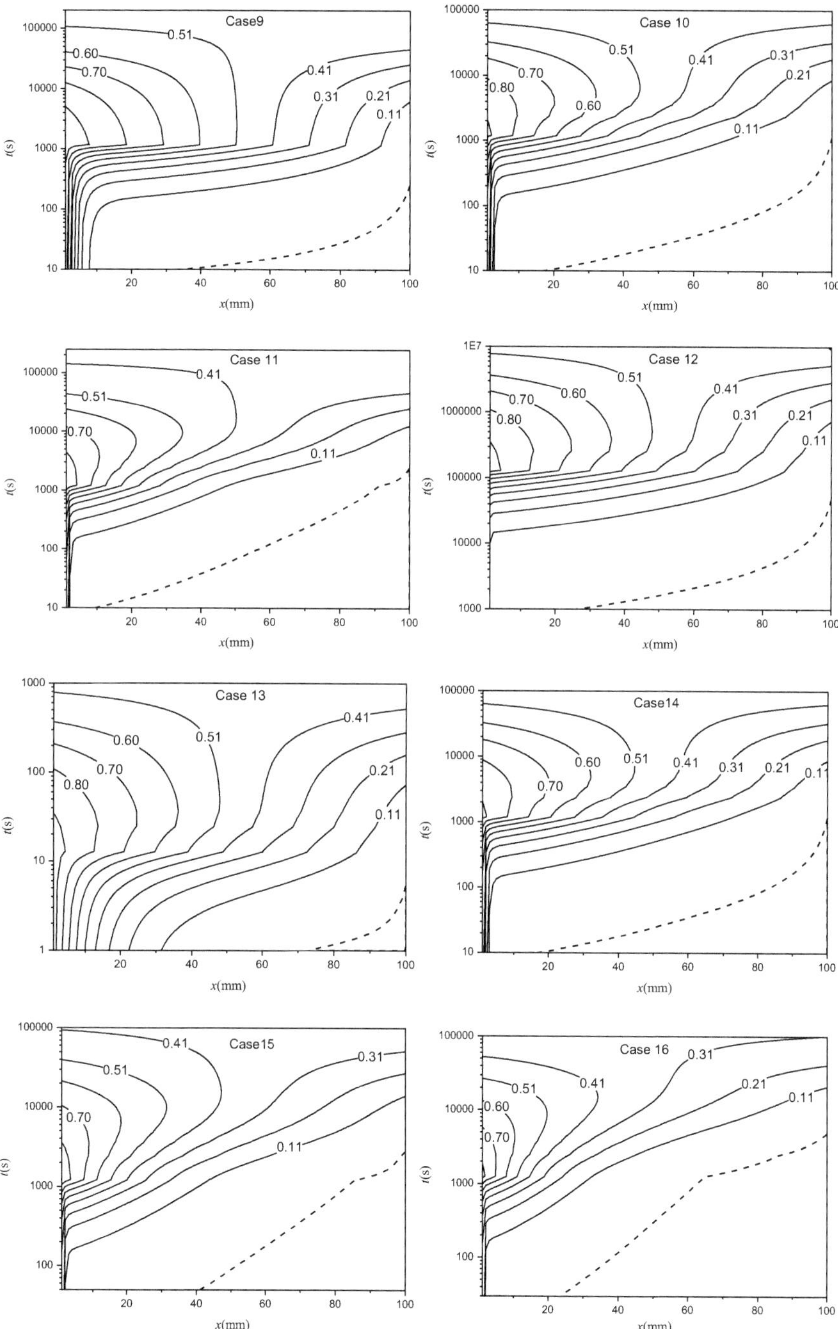

Fig. 3.5 (continued)

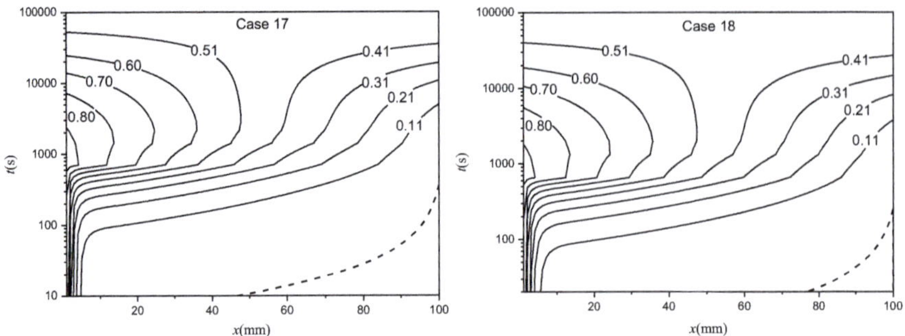

Fig. 3.5 (continued)

$t_D > 0.4$. Furthermore, we calculated the contribution η (defined as $\eta = S_1 / \sum_{m=1}^{\infty} S_m$, where S_m is the *m*th in Eq. (3.12)) of the first term in Eq. (3.12) to all terms when $t_D = 0.4$ at different values of *a* and *b*, as shown in Fig. 3.7. As can be seen from the figure, the first term (late time solution) contributes more than 99.7% for all cases when $t_D = 0.4$, indicating that the early time data can be neglected after $t_D > 0.4$. Therefore, $t_D > 0.4$ can be seen as a safe criterion for selecting the later time measurements. However, this criterion is not practical because the dimensionless time, t_D, contains the unknown permeability as defined in Eq. (3.14). We have to find a new criterion that is independent of permeability, which we will discuss next.

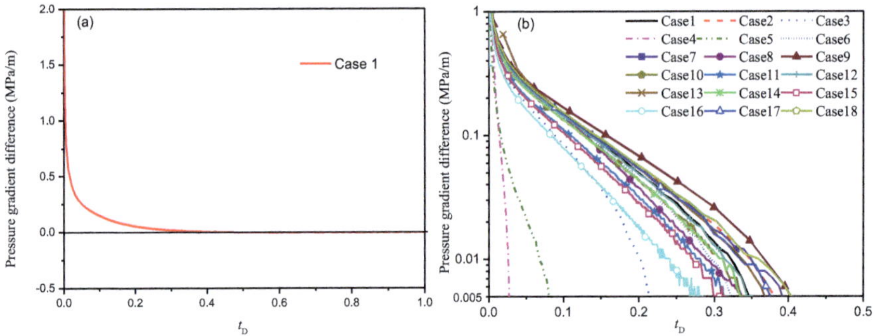

Fig. 3.6 The pressure gradient difference between two ends of the rock sample: **a** Case 1 and **b** Case 1 to Case 18

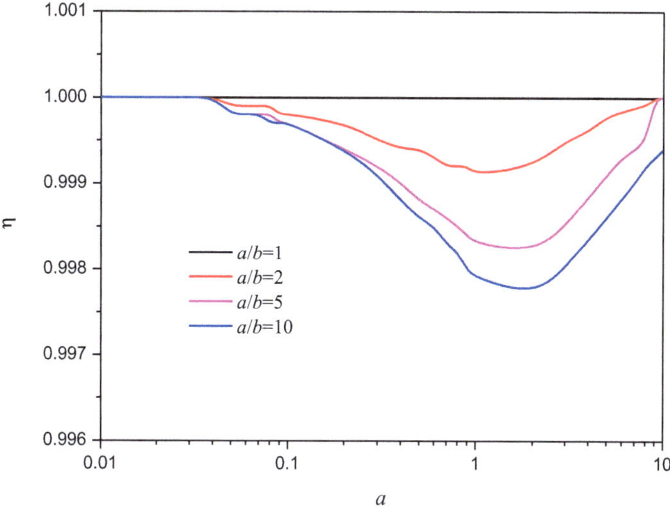

Fig. 3.7 Contribution of the first term in Eq. (3.12) to all terms when $t_D = 0.4$

3.4 A New Approach to Determine Permeability, Porosity and Adsorption Capacity

3.4.1 Determination of Apparent Porosity

As discussed earlier, the first step to determine accurate permeability from a pressure pulse decay experiment is to obtain the consistent apparent porosity which may vary with tested rock, type of working gas, and other experiment settings. We use the Boyle's law to estimate the apparent porosity of a rock sample from the mass balance equation at end of the given pressure pulse experiment.

$$\frac{V_u(p_0 + \Delta p)}{z_{p_0+\Delta p}} + \frac{(V_d + \phi_a AL)p_0}{z_{p_0}} = \frac{(V_u + V_d + \phi_a AL)p_f}{z_{p_f}} \tag{3.16}$$

where z is gas compressibility factor at given pressure (=1 for ideal gas) and p_f is the final pressure at end of pressure pulse experiment. A is the cross section area and L is the length of the rock sample. The apparent porosity can be easily solved from algebra Eq. (3.16) since all other parameters are known.

3.4.2 Determination of Apparent Permeability with Later Time Data

Although we understand that the time of the pressure pulse breakthrough is the turning point between the early time and later time pressure responses and that the pressure gradient within the rock samples becomes almost spatially independent after $t_D = 0.4$ (Fig. 3.6), the criterion in terms of t_D cannot be used directly to distinguish the early time and later time data in pressure decay method because t_D itself contains the unknown permeability which needs to be inferred from the measured data. To overcome this difficulty, inspired by Eq. (3.12), we propose to define a new scaled pressure decay, p^*, as follow:

$$p^* = \frac{1}{\theta_1^2} \ln \frac{2C_1}{\Delta p_D} \tag{3.17}$$

in which

$$C_1 = \frac{a\left(b^2 + \theta_1^2\right) + b\left[\left(a^2 + \theta_1^2\right)\left(b^2 + \theta_1^2\right)\right]^{0.5}}{\theta_1^4 + \theta_1^2\left(a + a^2 + b + b^2\right) + ab(a + b + ab)} \tag{3.18}$$

Considering that the first term of the general analytical solution Eq. (3.12) is a good approximation solution for later time as discussed before, we have the later time solution as below:

$$\Delta p_D = 2C_1 \exp\left(-t_D\theta_1^2\right) \tag{3.19}$$

By inserting Eqs. (3.17) into (3.19), we simply have:

$$p^* = t_D \tag{3.20}$$

In other words, the scaled decay curve, $p^*(t_D)$, is a 1:1 straight line in later time. Therefore, we can conveniently use $p^* > 0.4$ as the criterion to differentiate the early time data and late later data in pressure pulse tests. We re-plot the data of Figs. 3.4 in 3.8 in terms of p^* versus t_D. As shown in Fig. 3.8, although p^* can be positive or negative at early time depending on chamber volume ratio (e.g. negative for cases with $V_u > V_d$ and positive otherwise) all data are nicely scaled to the single 1:1 straight line after $t_D > 0.4$ (or $p^* > 0.4$). The neat thing is that p^* is independent of the unknown permeability. Therefore, the criterion base on p^* is more practical than that based on t_D.

Furthermore, by defining a scaled time,

$$t^* = \frac{t}{c\mu\phi_a L^2} \tag{3.21}$$

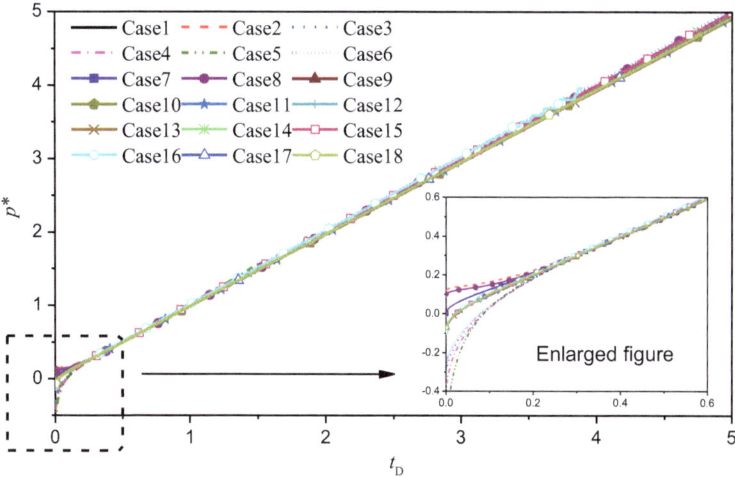

Fig. 3.8 The relationship of scaled pressure decay and dimensionless time for Case 1 to Case 18

we can rewrite Eqs. (3.20), using (3.14), as:

$$p^* = k_a t^* \tag{3.22}$$

Therefore, the apparent permeability can be calculated from Eq. (3.22). The procedure to process the measured data series $\{\Delta p_i, t_i\}$ can be summarized as follows:

Step 1 calculate apparent porosity by solving Eq. (3.16).
Step 2 calculate p_i^* using Eq. (3.17); t_i^* using Eq. (3.21).
Step 3 calculate the raw estimation of permeability, $k_i = \frac{p_i^*}{t_i^*} \quad if\ p_i^* > 0.4$
Step 4 calculate the mean and the standard deviation of estimated permeability $\{k_i\}$.

A list of source code implementing this algorithm in MATLAB is provided in Appendix. We note that there are other methods available for obtaining the permeability from the measured data series $\{\Delta p_i, t_i\}$. One such technique would be the linear regression. No attempts have been made to examine the rigor or nuances of those methods.

3.4.3 Adsorption and Slippage

The adsorption and slippage parameters can also be inferred from a set of pressure pulse decay experiments. First, the true porosity could be measured using non-adsorption gas or other independent porosity measurement methods under the same effective stress, then the equivalent porosity due to adsorption can be easily calculated using Eq. (3.6). Therefore, if the experiments are performed at two or more final pore pressures but the same effective stress, the common equilibrium adsorption models and corresponding parameters (e.g., Langmuir model, Freundlich model, D-A model, and BET model) can be deduced readily using Eq. (3.7).

After obtaining apparent permeabilities under two or more pore pressure with the same effective stress, the slippage parameter and true permeability can also be determined using Eq. (3.2).

3.4.4 Verification of the Proposed Method

To verify the accuracy of the proposed method, we calculated the parameters, using the method described in Sects. 3.4.2 and 3.4.3, from the simulated pressure decay data of 18 cases described in Sect. 3.3, and then compare them to the parameters actually used in the numerical simulations (Tables 3.3 and 3.4).

Table 3.3 shows the comparison between the calculated permeability using the proposed method from the data generated by numerical simulations (Case 1 through Case 13) and the true values used in numerical simulation. For comparison, we also included the estimated permeability from the same data sets using the methods from Brace et al. [11] and Cui et al. [23]. As shown in Table 3.3, the proposed method successfully estimated the permeability. The determined permeability values are very close to the true values. The relative error between the mean estimated value and the true value ranges from 0.1 to 1.2%. The calculated standard deviation is within 1.37% of the true value.

The calculated permeability using Cui's method is close to the true value for most cases. Since Cui's method doesn't differentiate the early-time and late-time data, large error can also be found in some cases, especially for cases with lager value of a, (e.g. the relative error is 8.8% in Case 4 for $a = 2$, and 25.1% in Case 5 for $a = 10$), in which the early time data deviates the single-exponential behavior significantly. An accurate estimation of permeability can improve the prediction of hydrocarbon in-place and recovery. Roadifer and Scheihing [30] have taken the example of an actual field reservoir and demonstrate that a permeability error of 20% can propagate throughout the model building and history matching process, and may cause 27.8% underestimation of oil recovery.

On the other hand, the permeability estimated by the Brace's method is close to the true value only if the volume ratios, a and b, are small (Case 9). When the volume ratios are large, the Brace's method can significantly underestimate permeability and the relative error can reach up to 72.5% (Case 5).

Table 3.4 shows the permeability results for the cases with adsorption effects. The permeability will be underestimated when testing sorptive rocks using Brace's solution even for cases with small volume ratios. Such trend is in qualitative agreement with the experimental results in Wang et al. [31] and Feng et al. [32]. These larger errors in estimated permeability are caused by two reasons: (1) violations of the strict assumptions used in developing Brace's solution; (2) error in porosity on permeability estimation (the apparent porosity rather than true porosity should be used). On the

Table 3.3 Calculated permeabilities and the true value (Case 1 to Case 13)

	True permeability ($10^{-18}m^2$)	Brace's method		Cui's method		The proposed method		
		k ($10^{-18}m^2$)	Relative error (%)	k ($10^{-18}m^2$)	Relative error (%)	k ($10^{-18}m^2$)	Relative error (%)	Standard deviation ($10^{-18}m^2$)
Case 1	1	0.981	−1.9	0.998	−0.2	0.997	−0.3	0.0035
Case 2	1	0.967	−3.3	0.994	−0.6	0.999	−0.1	0.0099
Case 3	1	0.815	−18.5	1.039	3.9	0.995	−0.5	0.0050
Case 4	1	0.674	−32.6	1.088	8.8	1.012	1.2	0.0063
Case 5	1	0.275	−72.5	1.251	25.1	1.007	0.7	0.0032
Case 6	1	0.975	−2.5	1.002	0.2	1.010	1.0	0.0137
Case 7	1	0.962	−3.8	0.978	−2.2	0.991	−0.9	0.0052
Case 8	1	0.766	−23.4	0.975	−2.5	0.998	−0.2	0.0041
Case 9	1	0.993	−0.7	0.996	−0.4	0.997	−0.3	0.0021
Case 10	1	0.964	−3.6	0.997	−0.3	0.997	−0.3	0.0022
Case 11	1	0.921	−7.9	0.998	−0.2	0.997	−0.3	0.0023
Case 12	0.01	0.981	−1.9	0.00998	−0.2	0.00997	−0.3	0.000027
Case 13	100	0.981	−1.87	99.8	−0.2	99.6	−0.4	0.539

other hand, the results of using Cui's method and our method match the true permeability very well, with maximum relative errors of 0.8% in our method and 0.9% in Cui's method. It is worth mentioning that in Cui's method the adsorption model and parameters are assumed to be known accurately, which requires additional adsorption experiments to obtain apparent porosity and only limits to Langmuir model. While for other adsorption models, permeability error can also be induced. For example, we carried out additional numerical cases using similar parameters as Case 14 but replacing Langmuir model with Freundlich model ($\psi(p) = 0.01p^{0.2}$) to generate pressure decay data. It turns out that a permeability error of 9.3% is produced by Cui's method. But only a permeability error of 0.6% is caused with our method. Note that, in our method, the apparent porosity is estimated based on Eq. (3.16) without assuming any particular adsorption model.

Table 3.4 Comparison between the calculated parameters and the true values in cases with adsorption

	True permeability (10^{-18}m^2)	Brace's method		Cui's method		The proposed method			
		k (10^{-18}m^2)	Relative error (%)	K (10^{-18}m^2)	Relative error (%)	K (10^{-18}m^2)	Relative error (%)	m_L	p_L (MPa)
Case 14	1	0.965	−3.5	0.997	−0.3	0.997	−0.3	0.0009	4.95
Case 15	1	0.906	−9.4	1.001	0.1	0.998	−0.2	0.0049	4.86
Case 16	1	0.841	−15.9	1.009	0.9	1.008	0.8	0.0099	4.82

Table 3.5 Permeability results of numerical tests with gas slippage effects

Numerical tests	True permeability ($10^{-18}m^2$)	Apparent permeability ($10^{-18}m^2$)	Absolute permeability ($10^{-18}m^2$)	Standard deviation ($10^{-18}m^2$)	λ
Case 17	1	1.4722	1.012	0.0073	0.48
Case 18	1	1.9473	1.013	0.0103	0.97

The adsorption parameters were calculated by solving Eq. (3.8b) using two apparent porosity values obtained from two sets of simulations with initial pressure at 1 MPa and 2 MPa, respectively, for Case 14–16. The calculated parameters match the true Langmuir parameters used in the numerical simulations very well (Table 3.4), supporting the usefulness of our method.

Table 3.5 shows the results of the cases with gas slippage effects. The slippage parameter λ and absolute permeability were calculated using Eq. (3.2) using the apparent permeabilities obtained from two sets of simulations of Case 17 and Case 18 with initial pressure at 1 MPa and 2 MPa, respectively. As listed in Table 3.5, the slippage parameter determined is nearly identical to the true values (Table 3.2). The apparent permeability increases with slippage effects for a given pressure.

To sum up, there are three main differences between our method and Brace's, Cui's method. First, the proposed method defines a new dimensionless pressure that allows consistent separation of the early and late time regimes, which is lack in Brace's and Cui's method. Second, in Cui's method, the apparent porosity is calculated with the known Langmuir isotherm model to remove the effects of the adsorption on the determination of the permeability. In our method, the apparent porosity is measured from the same pressure pulse tests using the initial pressure and final pressure, which is independent of particular adsorption models. Furthermore, with our method, we could not only remove the effects of the adsorption on the determination of the permeability but also could identify proper adsorption model and parameters with multiple pressure pulse tests. Third, the idea or method to simultaneously determine permeability, porosity, adsorption model/parameters, and slippage parameters from pressure pulse decay experiments is novel.

3.5 Application to Real Measurements

In this section, we applied the proposed method to process the real measured data under various experiment setting including different gas type. Two sets of pressure pulse decay experiments with different volume ratios were carried out. Test 1 was conducted on a sorptive rock, shale, at constant effective stress of 3 MPa by fixing the difference between the confining pressure (p_c) and the pore pressure (p). In this test, two types of gases (CH_4 and He) were used, respectively, among which CH_4 is found to be adsorbed by the shale. Test 2 was conducted on a rock-like sample (a mixture of ordinary Portland cement, sand, and water at a ratio of 1:1.2:0.65) at a constant pore pressure of 2 MPa but the confining pressures p_c ranges from 5 to 20 MPa, resulting in the effective stress varying from 3 to 18 MPa. Only He was used in Test 2. The purity of gases used in these experiments (He and CH_4) was 99.999%. The detailed experimental schemes are listed in Table 3.6.

The estimated permeability (with its standard deviation) and porosity of the samples are shown in Table 3.6. As expected, the apparent permeability in Test 1 is nearly identical for all cases since the effective stress is constant although the gas slippage effect slightly modifies the apparent permeability under the different pore pressure. The estimated slippage parameter λ is 0.34 for He and 0.20 for CH_4 and the true average permeability is $1.41 \times 10^{-18} \text{m}^2$. The apparent porosity in Test 1 with He shows little changes with the pore pressure (under the same effective stress), implying that He is not adsorbed by the shale. Therefore, the difference in the apparent porosity of the shale sample obtained with CH_4 and He can be seen as a measurement of the equivalent porosity due to adsorption of CH_4 by the shale. Given the relationship between the equivalent porosity due to adsorption and the associated pore pressure, we can easily obtain the adsorption model and corresponding parameters. Figure 3.9 shows the fitting curve of experimental data using different sorption models. The fitted parameters for these sorption models were listed in Table 3.7. Note that the experimental temperature were above methane's critical temperature (−82.7 °C), hence the pseudo-saturation vapor pressure [33, 34] was used and calculated in BET model and D-A model. The figure indicates that Langmuir modeled results show the best match with the experimental results (with R^2 value of 0.99). Compare to other sorption models, BET model has the poorest performance with R^2 smaller than 0.5. Moreover, independent adsorption experiments were conducted on shale sample to measure adsorption parameters, which gives the Langmuir parameters $m_L = 0.0014$ and $p_L = 5.67$ MPa, matching the calculated results (Table 3.7)

Table 3.6 Experimental schemes and results of Test 1 and Test 2

	Fluid	p_c (MPa)	p (MPa)	ϕ_a (%)	k_a (m^2)
Test 1	He	5	2	2.08	$1.594 \pm 0.044 \times 10^{-18}$
		6.5	3.5	2.01	$1.494 \pm 0.039 \times 10^{-18}$
		8	5	2.14	$1.447 \pm 0.038 \times 10^{-18}$
		9.5	6.5	1.89	$1.436 \pm 0.087 \times 10^{-18}$
	CH_4	5	2	6.02	$1.601 \pm 0.075 \times 10^{-18}$
		6.5	3.5	4.83	$1.557 \pm 0.068 \times 10^{-18}$
		8	5	4.05	$1.515 \pm 0.057 \times 10^{-18}$
		9.5	6.5	3.53	$1.498 \pm 0.080 \times 10^{-18}$
Test 2	He	5	2	24.03	$7.372 \pm 0.019 \times 10^{-17}$
		10	2	21.87	$4.525 \pm 0.036 \times 10^{-17}$
		15	2	20.94	$3.061 \pm 0.023 \times 10^{-17}$
		20	2	20.46	$2.306 \pm 0.041 \times 10^{-17}$

well. This proves that the proposed simultaneously determination of permeability, porosity, and adsorption capacity is effective.

Both the permeability and the porosity in Test 2 decrease with the increase of confining pressure for fixed pore pressure. This is consistent with the theory of the general relationship between the rock porosity/permeability and the effective stress. Because only one gas (non-adsorptive He) and a single pore pressure were used in Test 2, we could not infer the parameters of slippage and adsorption from these measurements.

Finally, with the obtained apparent porosities and permeabilities, the curves of p^* versus t_D are plotted in Fig. 3.10. The figure clearly illustrates that all data from these experiments (after $p^* > 0.4$) obey Eq. (3.20), the general later time scaled solution.

To validate the accuracy of the obtained parameters from experimental tests, these obtained parameters (permeability, porosity, adsorption and slippage parameters)

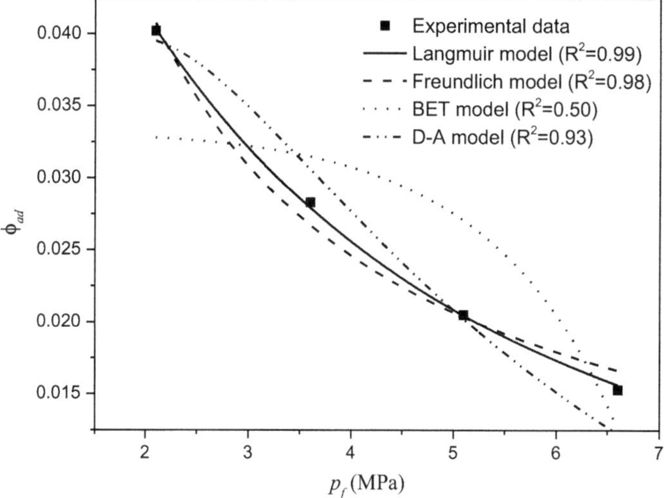

Fig. 3.9 The fitting curve of experimental data using different sorption models (Derivative function of sorption models are used to fitting experimental data)

Table 3.7 The modeled parameters for different sorption models

Sorption model	Function	Modeled parameters	R^2
Langmuir model	$\psi(p) = \frac{m_L p}{p_L + p}$	$m_L = 0.0011$; $p_L = 5.32$ MPa	0.99
Freundlich model	$\psi(p) = Kp^n$	$K = 8.62 \times 10^{-4}$; $n = 0.217$	0.98
BET model	$\frac{1}{\psi(p)(p_0/p-1)} = \frac{1}{V_m C} + \frac{C-1}{V_m C}\frac{p}{p_0}$	$V_m = 0.0013$; $C = 0.763$	0.50
D-A model	$\psi(p) = V_0 \exp\left[-D\left\{\ln\left(\frac{p_0}{p}\right)^2\right\}\right]$	$V_0 = 5.00 \times 10^{-4}$; $D = 0.439$	0.93

Note that p_0 is the saturation vapor pressure in BET model and D-A model

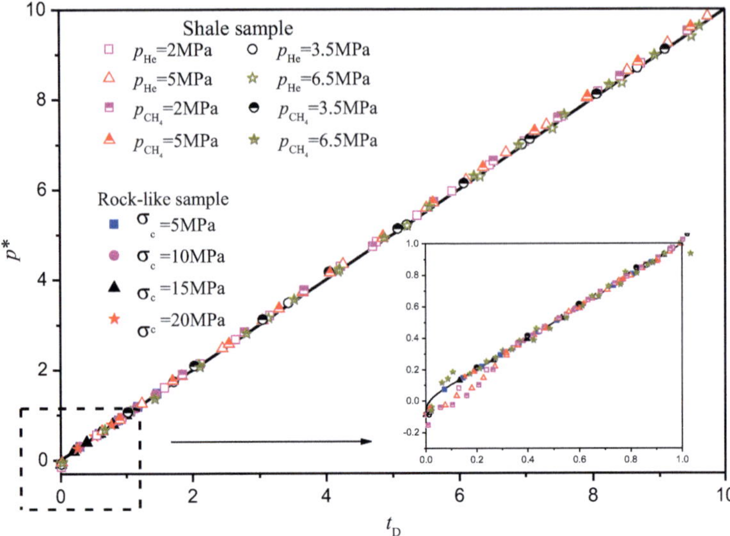

Fig. 3.10 The relationship of scaled pressure decay and dimensionless time for Test 1 and Test 2

were used as input for the numerical model to simulate the given pressure pulse decay experiments. Figure 3.11 shows the simulated upstream and downstream pressures compared against the measured values for one of the cases (CH_4 pressure of 2 MPa in Test 1) listed in Table 3.6 (others are similar). As can be seen from the figure, the solid curves match the solid circles very well, indicating that the parameters of the rock sample (permeability, porosity, adsorption and slippage parameter) obtained using the proposed method are correct.

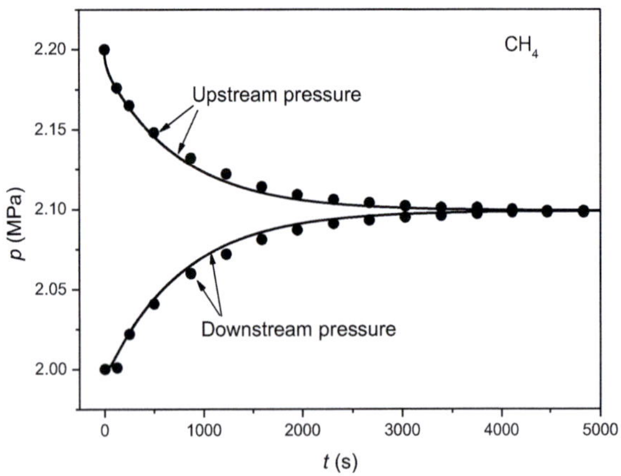

Fig. 3.11 Pressure responses in the upstream chamber and downstream chamber generated from numerical test (solid curves) and experimental test for the CH_4 pressure of 2 MPa in Test 1 (solid circles)

3.6 Conclusions

The pressure pulse decay method can be seen as an experiment of the pressure pulse propagation through the rock sample. The different flow conditions before and after the breakthrough of the pressure pulse result in different behaviors in terms of the observed pressure pulse decay curve, i.e., early time decay curve (before breakthrough) and later time decay curve (after breakthrough). With our new scaling method, the later time decay curves obtained from experiments with various rock properties and experiment setting parameters can be scaled into a single straight 1:1 line on a $p^*(t_D)$ plot (scaled decay vs. dimensionless time plot). It was found that $p^* > 0.4$ could be the universal criterion for determining the later time decay data used to estimate the permeability for all typical types of rock, gas, and experimental devices. Unlike previous similar criterion for determining later time decay data in literature, this criterion does not depend on the unknown permeability so that it can be easily used to automatically distinguish the later time data from the raw measurements.

Thoroughly analysis of the governing equations reveals that the permeability and the porosity controlling the pressure propagation in the rock are often apparent values if adsorption and gas slippage are existing (e.g., in shale gas applications). Therefore, the value directly estimated from pressure decay method is the apparent permeability while the apparent porosity (i.e., including the adsorption-induced storage) must be used in calculation of such apparent permeability. We proposed a method to estimate the apparent porosity based on the mass balance equation in terms of initial pressure and final pressure of the same pressure pulse decay experiment for determining permeability. The equivalent porosity of the rock sample due to adsorption can be further obtained if either the true porosity is known independently or a non-adsorptive gas is also used under the same experimental conditions. Furthermore, the adsorption model and corresponding parameters can be estimated from a set of pressure pulse decay experiments including different pore pressure but the same effective stress. In later case, the slippage parameter can also be inferred from the pressure pulse decay experiments.

Based on the findings above, a new robust approach to simultaneously determine permeability and porosity from pressure-pulse decay measurements is developed. The new approach is implemented into MATLAB to automatically process the pressure pulse decay data and estimate the apparent permeability including its standard deviation and the apparent porosity of rock samples. The tests of the method on the pressure pulse decay data obtained by numerical experiments using TOUGH + REALGASBRINE proves that the proposed approach can provide permeability, porosity, adsorption parameters, and slippage parameter that are very close to the true values (the input values used in numerical simulations) without limitation of particular adsorption models. We also demonstrated the successful applications of the proposed approach to the real experimental data involving two rock types, two gas types, and various pore and confining pressures.

Appendix

The main MATLAB code implemented for permeability and porosity calculation for pressure pulse decay test is shown as below.

```
clear all;clc;
[Vu,Vd,A,l,p0,pf,dp,u,c,z1,z2]=textread('parameter.txt','%f%f%f%f%f%f%f%f%f%f%f', 'delimiter', ',')
% please input the value of upstream and downstream chamber volume(Vu, Vd), rock size (A,l), initial pressure (p0), final equilibrium pressure (pf), pulse pressure (dp), fluid viscosity (u), fluid compressibility (c) and deviation factor of gas (z1 and z2) in a separated file named "parameter.txt". SI Units should be used.

Poro=-((Vu*dp+p0*(Vd+Vu))/z1-(pf*(Vd+Vu))/z2)/((A*l*p0)/z1-(A*l*pf)/z2); % porosity calculation
a=Poro*A*l/Vu % Volume ratio of pore to upstream
b=Poro*A*l/Vd % Volume ratio of pore to downstream


syms x
f1=tan(x)-(a+b)*x./(x.^2-a*b);
f11=matlabFunction(f1);
theta=fsolve(@(x)[f11(x)],1); % The first root of Eq.(3.13)
C1=(a*(b^2+(theta)^2)-(-
1)^1*b*sqrt((a^2+(theta)^2)*(b^2+(theta)^2)))/((theta)^4+(theta)^2*(a+a^2+b+b^2)+a*b*(a+b+a*b));

[t p]=textread('pressuredata.txt') %   please input measured data of time and differential pressure in a separated file named "pressuredata.txt"
pd=p/dp%   Dimensionless pressure
ps=1/theta^2*log(2*C1)-1/theta^2*log(pd) % scaled pressure decay
ts=t/l^2/(Poro)/u/c %

% permeability calculation
n=size(ps,1) % cycle index
j=1
for i=1:n
    if(ps(i)>0.4) % The early time data are skipped here.
        k(j)=ps(i)/ts(i)
        j=j+1
          end
end
km=mean(mean(k)) % avarage value permeability
kvar=var(k) % The variance of permeability
kstd=std(k) % The standard deviation of permeability
fprintf('Permeability= %e \n',km)
fprintf('Standard deviation of permeability= %e \n',kstd)
```

References

1. Karacan CÖ (2008) Evaluation of the relative importance of coal bed reservoir parameters for prediction of methane inflow rates during mining of longwall development entries. Comput Geosci 34(9):1093–1114. https://doi.org/10.1016/j.cageo.2007.04.008
2. Jin L, Hawthorne SB, Sorensen JA, Pekot LJ, Kurz BA, Smith SA, Heebink LV, Herdegen V, Bosshart NW, Torres Rivero JA, Dalkhaa C (2017) Advancing CO_2 enhanced oil recovery and storage in unconventional oil play-experimental studies on Bakken shales. Appl Energy 208:171–183. https://doi.org/10.1016/j.apenergy.2017.10.054
3. Fei Y, Johnson RL, Gonzalez M, Haghighi M, Pokalai K (2018) Experimental and numerical investigation into nano-stabilized foams in low permeability reservoir hydraulic fracturing applications. Fuel 213:133–143. https://doi.org/10.1016/j.fuel.2017.10.095
4. Pan L, Oldenbur CM, Freifeld BM, Jordan PD (2017) Modeling the Aliso Canyon underground gas storage well blowout and kill operations using the coupled well-reservoir simulator T2Well. J Petrol Sci Eng 161:158–174. https://doi.org/10.1016/j.petrol.2017.11.066
5. Kim TH, Cho J, Lee KS (2017) Evaluation of CO_2 injection in shale gas reservoirs with multicomponent transport and geomechanical effects. Appl Energy 190:1195–1206. https://doi.org/10.1016/j.apenergy.2017.01.047
6. Kim KY, Oh J, Han WS, Park KG, Shinn YJ, Park E (2018) Two-phase flow visualization under reservoir conditions for highly heterogeneous conglomerate rock: a core-scale study for geologic carbon storage. Sci Rep 8:1–10. https://doi.org/10.1038/s41598-018-23224-6
7. Berlepsch TV, Haverkamp B (2016) Salt as a host rock for the geological repository for nuclear waste. Elements 12(4):257–262. https://doi.org/10.2113/gselements.12.4.257
8. Nasir O, Nguyen TS, Barnichon JD, Alain M (2017) Simulation of the hydro-mechanical behaviour of bentonite seals for the containment of radioactive wastes. Can Geotech J 54(8):1055–1070. https://doi.org/10.1139/cgj-2016-0102
9. Zheng L, Rutqvist J, Xu H, Birkholzer JT (2017) Coupled THMC models for bentonite in an argillite repository for nuclear waste: illitization and its effect on swelling stress under high temperature. Eng Geol 230:118–129. https://doi.org/10.1016/j.enggeo.2017.10.002
10. Sander R, Pan Z, Connell LD (2017) Laboratory measurement of low permeability unconventional gas reservoir rocks: a review of experimental methods. J Nat Gas Sci Eng 37:248–279. https://doi.org/10.1016/j.jngse.2016.11.041
11. Brace WF, Walsh JB, Frangos WT (1968) Permeability of granite under high pressure. J Geophys Res 73:2225–2236. https://doi.org/10.1029/JB073i006p02225
12. Hsieh P, Tracy J, Neuzil C, Bredehoeft J, Silliman S (1981) A transient laboratory method for determining the hydraulic properties of 'tight' rocks-I theory. Int J Rock Mech Min Sci Geomech Abstr 18:245–252. https://doi.org/10.1016/0148-9062(81)90979-7
13. Dicker A, Smits R (1988) A practical approach for determining permeability from laboratory pressure-pulse decay measurements. In: International meeting on petroleum engineering. Society of Petroleum Engineers. https://doi.org/10.2118/17578-MS
14. Jones SC (1997) A technique for faster pulse-decay permeability measurements in tight rocks. SPE Form Eval 12:19–26. https://doi.org/10.2118/28450-PA
15. Metwally YM, Sondergeld CH (2011) Measuring low permeabilities of gas-sands and shales using a pressure transmission technique. Int J Rock Mech Min 48(7):1135–1144. https://doi.org/10.1016/j.ijrmms.2011.08.004
16. Yang Z, Sang Q, Dong M, Zhang S, Li Y, Gong H (2015) A modified pressure-pulse decay method for determining permeabilities of tight reservoir cores. J Nat Gas Sci Eng 27:236–246. https://doi.org/10.1016/j.jngse.2015.08.058
17. Hannon MJ (2016) Alternative approaches for transient-flow laboratory-scale permeametry. Transp Porous Med 114(3):719–746. https://doi.org/10.1007/s11242-016-0741-8
18. Feng R (2017) An optimized transient technique and flow modeling for laboratory permeability measurements of unconventional gas reservoirs with tight structure. J Nat Gas Sci Eng 46:603–614. https://doi.org/10.1016/j.jngse.2017.08.032

19. Chen T, Stagg PW (1984) Semilog analysis of the pulse-decay technique of permeability measurement. Soc Petrol Eng J 24:639–642. https://doi.org/10.2118/11818-PA
20. Kamath J, Boyer R, Nakagawa F (1992) Characterization of core scale heterogeneities using laboratory pressure transients. SPE Form Eval 7:219–227. https://doi.org/10.2118/20575-PA
21. He J, Ling K (2016) Measuring permeabilities of Middle-Bakken samples using three different methods. J Nat Gas Sci Eng 31:28–38. https://doi.org/10.1016/j.jngse.2016.03.007
22. Zhao Y, Zhang L, Wang W, Tang J, Lin H, Wan W (2017) Transient pulse test and morphological analysis of single rock fractures. Int J Rock Mech Min Sci 91:139–154. https://doi.org/10.1016/j.ijrmms.2016.11.016
23. Cui X, Bustin AMM, Bustin RM (2009) Measurements of gas permeability and diffusivity of tight reservoir rocks: different approaches and their applications. Geofluids 9:208–223. https://doi.org/10.1111/j.1468-8123.2009.00244.x
24. Feng R, Liu J, Harpalani S (2017) Optimized pressure pulse-decay method for laboratory estimation of gas permeability of sorptive reservoirs: part 1-background and numerical analysis. Fuel 191:555–564. https://doi.org/10.1016/j.fuel.2016.11.079
25. Klinkenberg LJ (1941) The permeability of porous media to liquids and gases. Paper presented at the API 11th mid-year meeting, Tulsa, Oklahoma, May 1941
26. Duong DD (1998) Adsorption analysis: equilibria and kinetics. Imperial College Press, London
27. Zhang H, Liu J, Elsworth D (2008) How sorption-induced matrix deformation affects gas flow in coal seams: a new FE model. Int J Rock Mech Min Sci 45(8):1226–1236. https://doi.org/10.1016/j.ijrmms.2007.11.007
28. Moridis GJ, Pruess K. (2014) User's manual of the TOUGH+ Core Code v1.5: a general-purpose simulator of non-isothermal flow and transport through porous and fractured media. Report LBNL-1871E, Lawrence Berkeley National Laboratory, Berkeley, CA
29. Ma T, Rutqvist J, Oldenburg CM, Liu W (2017) Coupled thermal-hydrological-mechanical modelling of CO_2-enhanced coalbed methane recovery. Int J Coal Geol 179:81–91. https://doi.org/10.1016/j.coal.2017.05.013
30. Roadifer RD, Scheihing MMH (2011) Impacts of the porosity permeability transform throughout the reservoir modelling workflow. In: SPE annual technical conference and exhibition, Denver Colorado, SPE 146574. https://doi.org/10.2118/146574-MS
31. Wang Y, Liu S, Elsworth D (2015) Laboratory investigations of gas flow behaviors in tight anthracite and evaluation of different pulse-decay methods on permeability estimation. Int J Coal Geol 149:118–128. https://doi.org/10.1016/j.coal.2015.07.009
32. Feng R, Harpalani S, Pandey R (2016) Evaluation of various pulse-decay laboratory permeability measurement techniques for highly stressed coals. Rock Mech Rock Eng 50:297–308. https://doi.org/10.1007/s00603-016-1109-7
33. Agarwal RK, Schwartz JA (1988) Analysis of high-pressure adsorption of gases on activated carbon by potential theory. Carbon 26(6):873–887. https://doi.org/10.1016/0008-6223(88)90111-X
34. Harpalani S, Prusty BK, Dutta P (2006) Methane/CO_2 sorption modeling for coalbed methane production and CO_2 sequestration. Energy Fuels 20(4):1591–1599. https://doi.org/10.1021/ef0504341

Chapter 4
A Modified One-Chamber Pressure-Pulse Decay Method

4.1 Introduction

Permeability and adsorption capacity are two critical geological parameters governing the productivity of shale gas wells and the effectiveness of CO_2 geological storage. In shale reservoirs, gas is stored both as free gas within matrix pores and natural fractures, and as adsorbed gas on the internal surfaces of organic matter and clay minerals [1]. The adsorption of CH_4 significantly influences the permeability evolution of unconventional reservoirs, as it induces matrix swelling and consequently reduces permeability. This phenomenon represents a major challenge in practical field applications [2–5].

Over the past two decades, laboratory studies have extensively examined the effect of adsorption-induced swelling on permeability. For instance, Zhang et al. [6] observed that adsorption leads to microstructural alterations in coal, substantially reducing its permeability. Notably, for CO_2, adsorption-induced pore structure modifications are often irreversible during injection and depletion cycles. Given that different gases cause varying degrees of swelling in coal and shale, numerous studies have focused on sorption-induced deformation of CO_2, CH_4, N_2 [7–10], and their mixtures [11–13]. These investigations further indicate that coal and shale exhibit anisotropic swelling behavior, with sorption-induced strain typically greater perpendicular to the bedding plane [14–16]. Based on experimental swelling data, several predictive models have been developed, including linear relationships [17], Langmuir-type equations [18, 19], and energy balance approaches [20]. Subsequently, numerous permeability models incorporating these swelling behaviors have been proposed to address the impact of adsorption-induced swelling [21–26].

Despite progress in understanding adsorption-permeability interactions through experimental studies, two key limitations remain. First, conventional permeability measurements often overlook the influence of adsorption on gas storage [27–30], potentially leading to unreliable results. As highlighted by Cui et al. [24], ignoring adsorption effects can result in severe underestimation of permeability—exceeding

C. Wang and Y. Zhao, *Permeability Measurement of Tight Rock*,
https://doi.org/10.1007/978-981-92-0597-4_4

60% under conditions of low pressure and strong adsorption. Similarly, Cao et al. [31] reported a 97% error in shale permeability measurements when adsorption was disregarded, while Feng et al. [32] found that permeability corrected for adsorption effects was 52% higher than uncorrected values in coal. To date, few experimental methods accurately determine permeability under adsorption-influenced conditions. Cui et al. [24] integrated a Langmuir model into the pressure-pulse decay method, but their approach requires separate adsorption and permeability tests, which is time-consuming. Feng et al. [33, 34] proposed a bidirectional pressure-pulse test to mitigate adsorption effects; however, adsorption and desorption during pressure decay may not be fully balanced, even when initial and final equilibrium pressures are identical.

Second, methods for the continuous and simultaneous measurement of permeability, adsorption, and swelling behavior remain underdeveloped [35]. Most adsorption experiments utilize crushed samples to reduce testing time, but this approach cannot replicate in situ conditions, fails to account for natural fracture networks, and is influenced by particle size effects [37]. Although the Gas Research Institute (GRI) technique [36] allows permeability measurements on crushed samples as part of adsorption experiments, it is not applicable under reservoir conditions and does not reflect the flow characteristics of intact rock.

To address these gaps, this section presents an experimental method designed to simultaneously measure permeability and adsorption parameters. The proposed approach accounts for both adsorption storage effects and swelling-induced permeability changes, providing an efficient and integrated means to characterize adsorption-permeability relationships under representative conditions.

4.2 Experiment

4.2.1 Rock Sample

Two core samples, designated as Sample 1 and Sample 2, were obtained from an outcrop of the Lower Silurian Longmaxi Formation in the eastern Chongqing region of the Sichuan Basin, China. For the adsorption-permeability experiments, cylindrical specimens measuring 50 mm in diameter and 100 mm in length were prepared, as illustrated in Fig. 4.1. The study focuses on single-phase gas flow. Prior to testing, the shale samples were dehydrated by placing them in a vacuum drying oven at 70 °C for a minimum of 24 h. A suite of shale characterization analyses was performed to determine key petrophysical and geochemical properties, including mineral composition, total organic carbon (TOC) content, vitrinite reflectance (Ro), and porosity. The physical and chemical properties of the shale samples are summarized in Table 4.1.

Fig. 4.1 Image of shale samples

Table 4.1 Physical and chemical properties of the shale sample

Ro (%)	TOC (%)	Mineral composition (wt %)						
		Major (>25%)	Minor (5–25%)				Trace (<5%)	
3.72	2.88	Quartz	Potassium feldspar	Albite	Illite	Chlorite	Calcite	Pyrite

4.2.2 *Adsorption-Permeability Experiment*

This section introduces a novel method for the simultaneous measurement of rock adsorption capacity and permeability. The approach integrates a standard adsorption test with a modified pressure-pulse decay (PPD) experimental procedure. A schematic diagram of the experimental setup is provided in Fig. 4.2. The apparatus comprises a reference chamber, a sample cell, a high-resolution pressure transducer, a temperature-regulated air bath, a loading unit for applying confining and axial pressures, and a strain measurement unit. The procedural workflow for the adsorption-permeability experiment is illustrated in Fig. 4.3.

The step-by-step procedure for the simultaneous measurement is as follows:

1. Put the core sample into the sample cell, and load axial and confining pressure on the core sample. Prior to the adsorption-permeability experiment, a leak test of the apparatus is tested at the maximum applied pressure with Helium. Then the

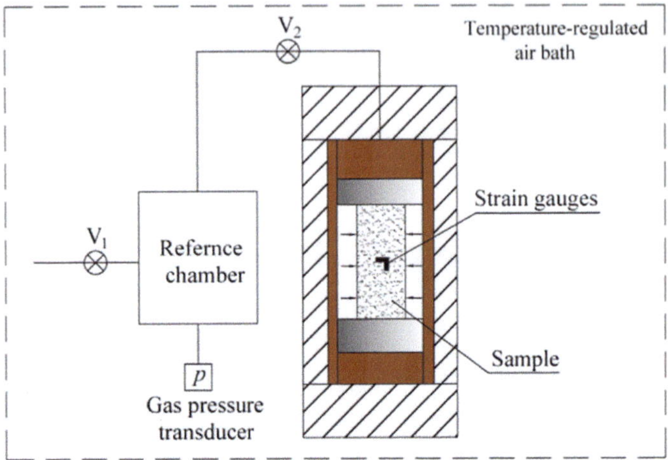

Fig. 4.2 Schematic diagram of essential part of the experimental setup

volume calibration experiments are performed to obtain the reference chamber volume and void volume.

2. The system is vacuumed for at least one day to exclude the influence of other gases on adsorption-permeability experiment. Then the initial equilibrium temperature T and CH_4 pressure p_{s0} of the system are set and recorded.
3. Close the sample cell isolation valve (V_2) and then charge the reference chamber an appropriate amount of CH_4 and record the equilibrium pressure (p_{c0}).
4. Open the isolation valve V_2 and record the pressure response with time until final equilibrium pressure of the system is attained. During this process, the adsorption-induced swelling is also measured by two linear deformation strain gauges (in axial and radial directions).
5. Close the sample cell isolation valve and carry out the next pressure point experiment. The temperature in the experiments is set to 40 °C (313.15K), and the pore pressure ranges from 0.5 to 7MPa. To eliminate the mechanical compression effect, the effective stress remains at a constant value (5MPa) in all adsorption-permeability experiments by fixing the difference between the confining pressure and the pore pressure.

The above experiment method involves two flow processes: pressure pulse propagation from the reference chamber to rock sample and CH_4 adsorbed on rock matrix surface. Therefore, permeability can be calculated from the pressure pulse decay data and CH_4 adsorption amount can be calculated from the mass change in terms of initial pressure and final pressure. The calculation method for simultaneously determining permeability and adsorption amount is introduced as followings.

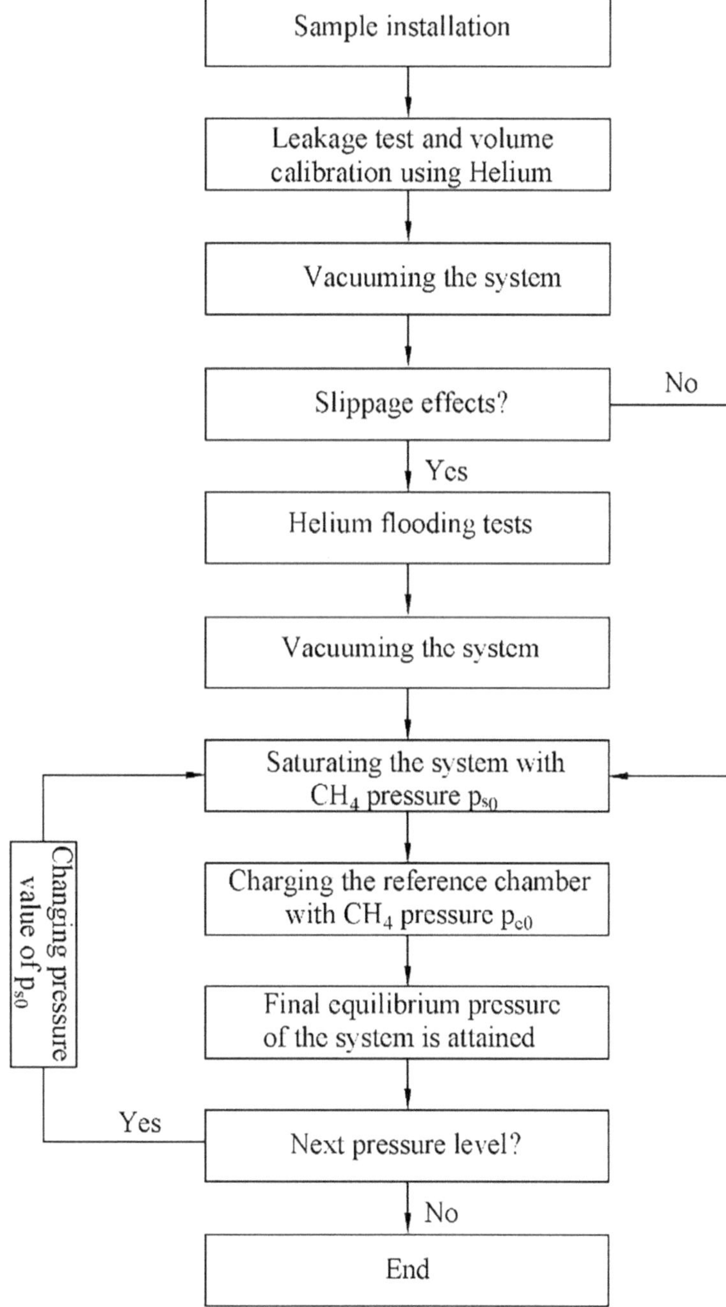

Fig. 4.3 Flow chart of adsorption-permeability experiments

4.2.3 Determination of Adsorption Amount

The volumetric method was used to determine the adsorption amounts of methane. The experimental adsorption amount, i.e., Gibbsian surface excess (GSE) of methane was calculated by [38–40]

$$\Delta GSE = \frac{1}{mRT}\left(\frac{p_{c0}V_c}{z_2} + \frac{p_{s0}V_v}{z_1} - \frac{p_f V_c}{z_3} - \frac{p_f V_v}{z_3}\right) \tag{4.1}$$

in which m donates sample mass, R is the universal gas constant, T is isothermal temperature; p_f is final equilibrium pressure of the system. V_c is the volume of the reference chamber, V_v is the void volume; z_1 z_2 and z_3 are the compressibility factor of methane at p_{s0}, p_{c0}, and p_f, respectively.

With the obtained excess adsorption amount (GSE), the absolute adsorption amount n_a can be corrected as [41, 42]

$$n_a = GSE\left(1 - \frac{\rho_g}{\rho_a}\right)^{-1} \tag{4.2}$$

where ρ_g and ρ_a are gas density in free phase and adsorbed phase, respectively.

4.2.4 Permeability Determination

A similar experimental design (in which only one chamber is used in pulse pressure tests to measure permeability) and derivation process for the analytical solution of permeability has been developed by Yang et al. [43]. They suggested that the measured results from the single-chamber pulse pressure decay method are much more accurate than that from the traditional pulse pressure decay method [44], and this modified method needs fewer transducers, a simpler experimental apparatus, and easier operation. However, their method is not suitable for testing sorptive rocks. In order to overcome this, we integrate the adsorption effects into Yang's method.

Assuming that Darcy's law prevails, the fundamental process of the aforementioned experiment can be seen as one-dimensional gas flow through the porous core sample. The governing equation is written as:

$$\frac{\partial M}{\partial t} = \frac{\partial}{\partial x} \cdot \left(\frac{k\rho}{\mu}\frac{\partial p}{\partial x}\right) \tag{4.3}$$

where M is the total gas mass in the control volume; ρ is the density of gas; t is time; μ is gas viscosity; k is permeability. The mass M consists of free-phase gas and adsorbed gas, and is described as followings [45–47]:

$$M = \phi\rho + (1-\phi)\rho_R\rho_{gs}\frac{pV_L}{p+p_L} \tag{4.4}$$

where ρ_R is the density of the rock sample, ρ_{gs} is the gas density at standard (atmospheric pressure and 273.15 K temperature) condition, V_L and p_L are Langmuir volume and pressure constants respectively.

Substituting Eqs. (4.4) into (4.3), we obtain,

$$\left[\phi\frac{d\rho}{dp} + \left(\rho - \rho_R\rho_{gs}\frac{pV_L}{p+p_L}\right)\frac{d\phi}{dp} + (1-\phi)\rho_R\rho_{gs}\frac{V_L p_L}{(p+p_L)^2}\right]\frac{\partial p}{\partial t} = k\frac{\rho}{\mu}\frac{\partial^2 p}{\partial x^2} \tag{4.5}$$

Assuming that $\frac{d\phi}{dp} = 0$ prevails because of a small pressure pulse in adsorption-permeability test. Following Cui's method, an apparent porosity is defined as [24]:

$$\phi_a = \phi + \phi_{ad} = \phi + \frac{\rho_R\rho_{gs}(1-\phi)}{c\rho}\frac{V_L p_L}{(p+p_L)^2} \tag{4.6}$$

in which ϕ_{ad} is the porosity increment induced by adsorption, V_L and p_L are Langmuir parameters, which can be simultaneously estimated through this experiment, as illustrated in the next section, and c is the gas compressibility,

$$c = \frac{1}{\rho}\frac{d\rho}{dp} \tag{4.7}$$

The gas density ρ is given as

$$\rho = \frac{p}{zRT} \tag{4.8}$$

where z is the gas compressibility factor, R is the universal gas constant, and T is temperature.

Substituting Eqs. (4.6), (4.7), and (4.8) into (4.5), the governing equation can be simplified as

$$\frac{\partial\rho(x,t)}{\partial t} = K\frac{\partial\rho^2(x,t)}{\partial x^2} \tag{4.9}$$

where K is the effective gas diffusivity (Cui et al. 2009),

$$K = \frac{k}{c\mu\phi_a} \tag{4.10}$$

Finally, the gas flow process in the pressure pulse decay method with single chamber can be described by the governing Eq. (4.9) completed by the following initial and boundary conditions [43]:

$$\lambda L\frac{d\rho_c}{dt} = K\frac{\partial\rho}{\partial x}|_{x=0} \tag{4.11a}$$

$$\frac{\partial\rho}{\partial x}|_{x=L} = 0 \tag{4.11b}$$

$$\rho(0,t) = \rho_c(t) \tag{4.11c}$$

$$\rho_c|_{t=0} = \rho_{c0} \tag{4.11d}$$

$$\rho|_{t=0} = \rho_{s0} \tag{4.11e}$$

where λ is the ratio of the chamber volume to the pore volume; L is rock sample length; ρ_c is the gas density of the chamber; ρ_{cO} is the initial gas density in the chamber, and ρ_{sO} is the initial gas density in the core sample.

Note that Eq. (4.9) have the same mathematical forms of the governing equations presented by Yang et al. [48]. Therefore, the Yang's late time solution can be readily presented by replacing porosity with apparent porosity [43].

$$k = -\frac{L^2\phi_a\mu cs}{\theta_1^2} \tag{4.12}$$

where θ_1 is the first root of the equation: $\tan\theta = -\lambda\theta$; s is the slope of straight line $\ln(F_R)$ versus t, where F_R, a dimensionless parameter, is gas residual ratio of remaining gas over total gas to be taken up by sample, which can be obtained from the pressure decay data (p_c) of the reference chamber [43].

$$F_R=1 - \frac{(p_{c0} - p_c)(1+\lambda)}{p_{c0} - p_{s0}} \tag{4.13}$$

where p_c is the gas pressure of the chamber; p_{cO} is the initial pressure in the chamber, and p_{sO} is the initial pressure in the core sample. Note that the apparent porosity is used in Eq. (4.12) to account for the contribution of CH_4 adsorption storage.

Moreover, gas permeability in tight rocks can be affected by the slip phenomenon due to the interaction between gas molecules and the flow path wall [49]. The gas slippage phenomenon can be mathematically described as:

$$k = k_\infty\left(1+\frac{b}{p}\right) \tag{4.14}$$

where k is apparent permeability measured from the adsorption-permeability test, k_∞ is intrinsic permeability, and b is the Klinkenberg coefficient, which can be obtained from Helium flooding experiments. When the shale sample is flooded with CH_4, both the slippage effect and adsorption effect govern the total permeability response.

However, when the shale is exposed only to Helium, the slippage effect alone takes place. Therefore, we can obtain the Klinkenberg coefficient from Helium flooding experiments.

4.3 Results and Discussion

4.3.1 Adsorption and Strain Results

Figure 4.4 shows the absolute adsorption isotherm of shale samples exposed to CH_4. Obviously, the absolute adsorption amount increases with pressure. Langmuir isotherm model is applied to describe the adsorption data and the corresponding fitting curves are also plotted in Fig. 4.4. The figure illustrates that the Langmuir model can describe the CH_4 adsorption on shale very well.

The adsorption-induced swelling strain was measured for each pressure step. Figure 4.5 shows the adsorption-induced strain with increasing CH_4 pressure. As can be seen from the figure, swelling strain from CH_4 adsorption first shows a rapid initial rise, followed by a gradual increase with time. These strain data can be well captured by the empirical Langmuir-type fit, which is consistent with the experimental results from Liu et al. [50], Lin et al. [51], and Chen et al. [52]. The mechanism of adsorption-induced swelling is presented as follows.

When shale micropores are filled with adsorptive gas (e.g. CH_4), the mean distances between sorbate molecules decrease. Hence, intermolecular repulsion forces between the sorbate-sorbate and sorbate-sorbent molecules should come into play. This leads to rising pressure inside the micropores and results in the expansion of the sorbent matrix. In the early stage of adsorption, the adsorption sites were occupied by the first layer of sorbate molecular. Both the intermolecular repulsion forces between the sorbate-sorbate and sorbate-sorbent molecules lead to pressure increase inside the pores, so the swelling strain increases rapidly at early adsorption stage. With the adsorption amount increasing, the adsorption sites almost had been occupied by the first layer sorbate molecular, therefore, intermolecular repulsion forces between sorbate-sorbent molecular make fewer contributions to the expansion of the sorbent matrix.

Furthermore, we plot the relationship between adsorption-induced swelling strain and the amount of gas adsorbed for CH_4 in Fig. 4.6. As can be seen from Fig. 4.6, the adsorption-induced swelling is almost linear to the adsorption amount, which is also observed by other authors [53, 54].

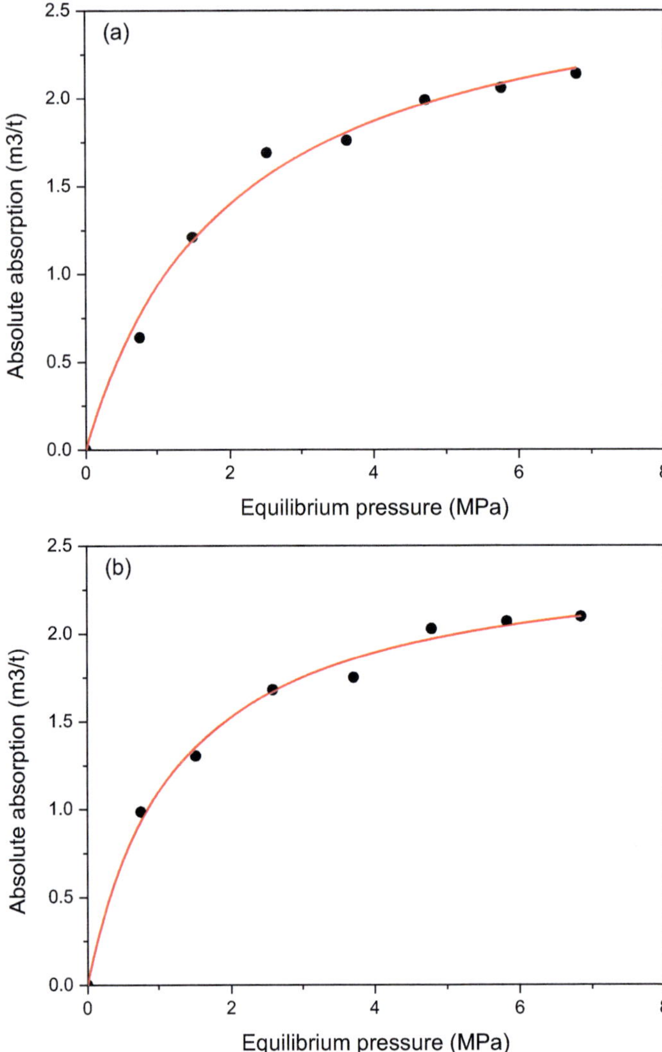

Fig. 4.4 CH_4 adsorption isotherm (solid circles) and Langmuir model fit (solid curve) of: **a** Sample 1 and **b** Sample 2

4.3.2 Permeability Results

The first step to determine permeability from the adsorption-permeability experiment is to obtain the apparent porosity through Eq. (4.6). Then the measured pressure decay versus time curve as well as the coefficient F_R versus time curve are plotted in Fig. 4.7. As the curves at different pressure steps are similar, only the results for pressure of 1MPa for Sample 1 are illustrated. The late-time data of F_R versus time are fitted using

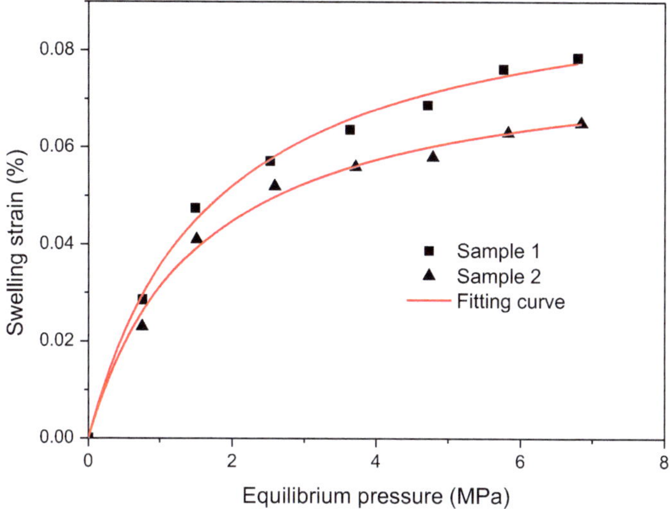

Fig. 4.5 Adsorption-induced strain and Langmuir-type curve fit

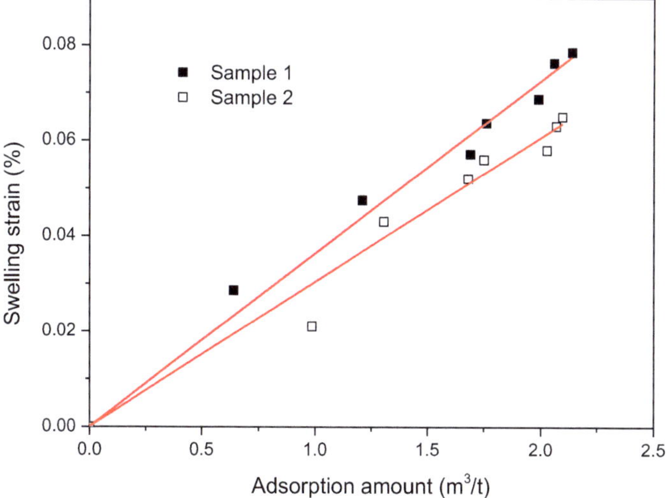

Fig. 4.6 The relationship of swelling strain and CH_4 adsorption amount

a linear function to obtain the parameter s in Eq. (4.12). Gas properties are calculated from the National Institute of Standards and Technology (NIST) webbook at http://webbook.nist.gov/chemistry/fluid/. The permeability, therefore, can be calculated after all the unknown parameters in Eq. (4.12) have been obtained.

The calculated permeability data from the above experiments were further reduced by the Klinkenberg correction to the intrinsic permeability. Figure 4.8 shows the

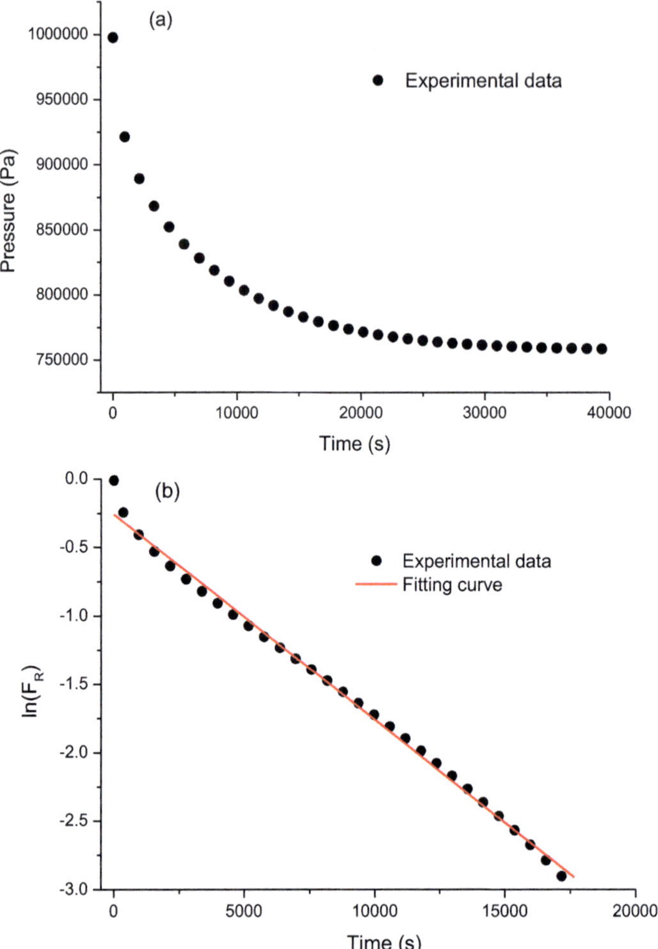

Fig. 4.7 Experimental results for permeability measurement of Sample 1 under pressure of 1 MPa: **a** pressure decay versus time. **b** $\ln(F_R)$ versus time

evolution of intrinsic permeability with pressure. With a constant effective stress condition, the intrinsic permeability is drastically reduced in the adsorption process. For example, the intrinsic permeability of Sample 1 decreases from $9.66 \times 10^{-19}m^2$ to $3.24 \times 10^{-19}m^2$ as the pressure increases from 0.76MPa MPa to 6.79MPa. Such a large permeability reduction is consistent with previous studies: Zhang et al. [6] measured a two order magnitude permeability reduction in a CO_2 flooding test, and Siriwardane et al. [55]measured a 90% permeability decrease in coal after CO_2 exposure. This can be contributed to the adsorption-induced matrix swelling which leads to natural fractures closure and hence reducing permeability.

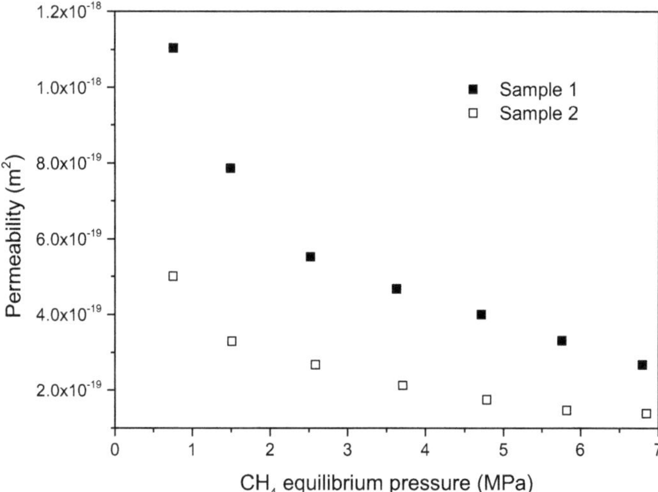

Fig. 4.8 The relationship of permeability and equilibrium pressure

4.3.3 Effects of Adsorption on Permeability

The effects of adsorption on permeability are influenced by two factors: 1) adsorption-induced storage which causes an incremental in apparent porosity, leading to a significant error in permeability measurement if true porosity is used; 2) adsorption-induced swelling which potentially closes the existing natural fractures and reduces the intrinsic permeability [56].

(1) Adsorption-induced storage

Close analysis of the governing equation of the adsorption-permeability experiment reveals that the porosity used in calculations should be the apparent porosity including all compression storage such as adsorption-induced storage. Hence, errors in the porosity (if true porosity is used) could propagate into the estimated permeability. To show this, we present the estimated permeability results of adsorption-permeability experiments using true porosity and apparent porosity in Fig. 4.9. As shown in Fig. 4.9, the permeability will be underestimated if true porosity is used. The permeability error increases with the decrease of pore pressure, which can be attributed to the more significant adsorption effect at low pressure, leading to a large increment in apparent porosity (Eq. 4.6). Similar experimental results are also observed by Wang et al. [57] and Feng et al. [32] through separate permeability and adsorption tests.

The adsorption storage effects also depend on the experiment configurations (the volume ratio of chamber volume over pore volume). We define the growth rate of permeability resulting from adsorption storage effects as

$$\eta = \frac{k_{ap} - k_{tp}}{k_{tp}} \times 100\% \tag{4.15}$$

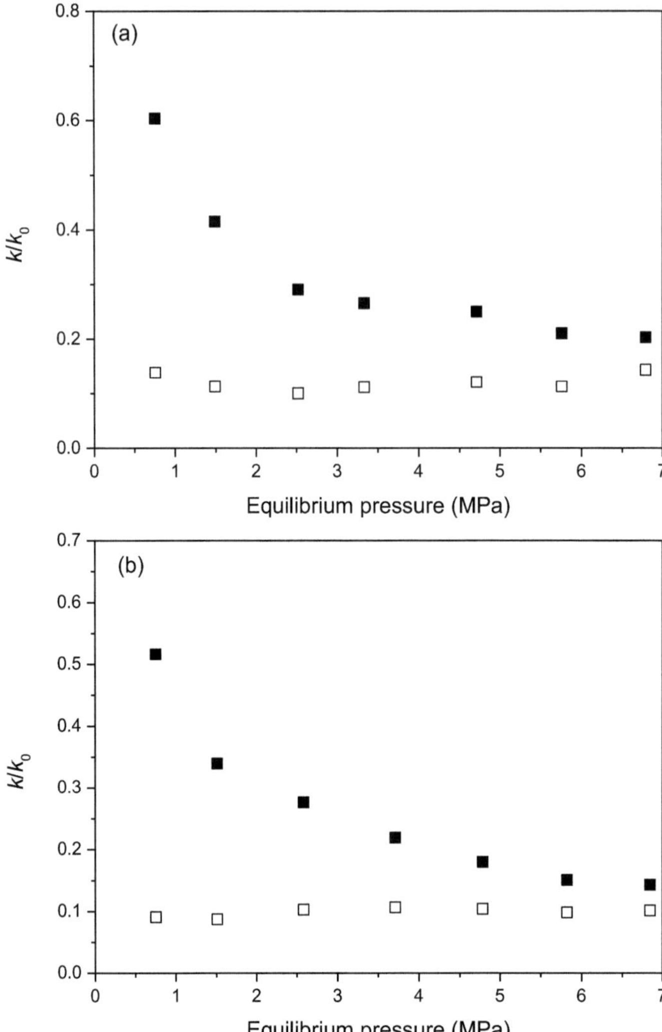

Fig. 4.9 Comparison between calculated permeabilities using true porosity (empty symbol) and apparent porosity (solid symbol) for **a** Sample 1 and **b** Sample 2

where k_{ap} and k_{tp} are the permeabilities when apparent porosity and true porosity are used in calculations, respectively. Figure 4.10 shows the evolution of the growth rate of estimated permeability as a function of the volume ratio. As shown in Fig. 4.10, the permeability growth rate is sensitive to the ratio of chamber volume over the pore volume. It first increases when λ is lower than a certain value (about 0.75), and then decreases monotonously. In other words, the adsorption storage effect exhibits a maximum point when the volume ratio is 0.75.

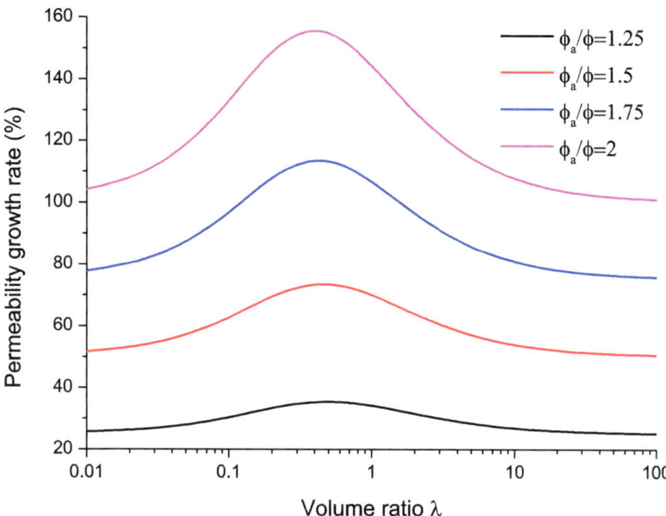

Fig. 4.10 Permeability growth rate versus λ for different ratio of apparent porosity over true porosity

(2) Adsorption-induced swelling

CH_4 adsorption in the pores is also expected to lead to pore structure modification and to correspondingly influence the intrinsic permeability in shale sample. The direct evidence of swelling-induced changes in coal microstructure due to gas adsorption is observed by Zhang et al. [58]. In their experiments, permeability loss was caused by closure of 80% of the micro fractures due to matrix swelling. Therefore, the permeability can be related to its porosity through

$$\frac{k}{k_0} = \left(\frac{\phi_0 - f(\varepsilon_V)}{\phi_0}\right)^3 \tag{4.16}$$

where k_0 is the initial permeability at initial porosity ϕ_0, ε_V is the adsorption volumetric strain, $f(\varepsilon_V)$ is a function describing the change in porosity induced by adsorption swelling.

The experimental results of intrinsic permeability versus adsorption-induced swelling strain are shown in Fig. 4.11. Permeability consistently dropped rapidly with the increase of swelling strain. This permeability drop can be fitted with a cubic curve by assuming $f(\varepsilon_V)$ as a simple linear function (printed on the graphs in Fig. 4.11).

$$\frac{k}{k_0} = (1 - a\varepsilon_V)^3 \tag{4.17}$$

where a is the coefficient which relies on the porosity, compressibility, Young's modulus, and Poisson's ratio. For the samples tested, the values of a are 5.469 for

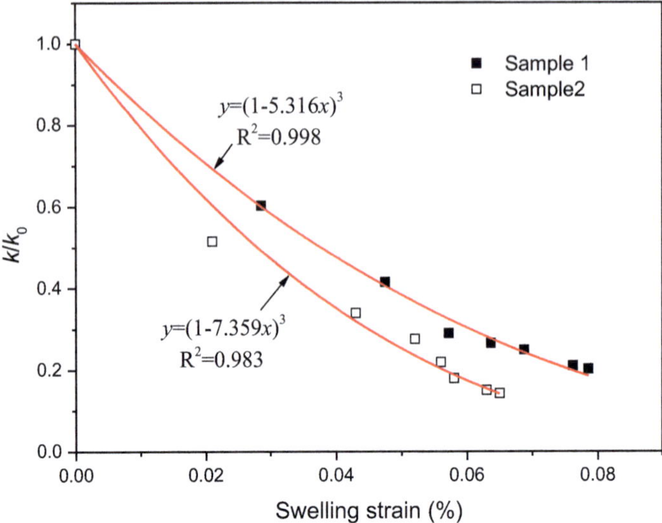

Fig. 4.11 The relationship of intrinsic permeability and adsorption-induced swelling strain

Sample 1 and 7.359 for Sample 2. It is worth mentioning that a linear relationship between the sorption strain and the amount of CH_4 adsorbed is found in this research and other researches [53, 54]. Hence the relation between permeability and the amount of gas adsorbed can also be described through a cubic curve.

4.4 Conclusion

(1) A new methodology was developed to simultaneously measure adsorption and permeability of reservoir rocks. The proposed method offers intrinsically consistent adsorption-permeability data to characterize unconventional reservoirs.
(2) The effects of adsorption on permeability are influenced by two factors: adsorption-induced storage and adsorption-induced swelling. The adsorption storage effects have a significant influence on permeability measurement, especially at low pressure. Meanwhile, the adsorption storage effect exhibits a maximum effect on permeability when the volume ratio is 0.75. To eliminate the adsorption-induced storage effect on permeability measurement, apparent porosity should be used.
(3) The CH_4 adsorption induced swelling may modify the pore structure in shale, as a consequence, permeability is reduced. The change in permeability associated with adsorption-induced swelling was found to be a cubic function with the swelling strain. Since the swelling strain is linearly proportional to the amount

of CH_4 adsorbed, the behaviors of permeability and the amount of adsorbing gas follow similar trends.

References

1. Yuan W, Pan Z, Li X, Yang Y, Zhao C, Connell LD, Li S, He J (2014) Experimental study and modelling of methane adsorption and diffusion in shale. Fuel 117:509–519
2. Kiyama T, Nishimoto S, Fujioka M, Xue Z, Ishijima Y, Pan Z, Connell LD (2011) Coal swelling strain and permeability change with injecting liquid/supercritical CO_2 and N_2 at stress-constrained conditions. Int J Coal Geol 85:56–64
3. Perera MSA, Ranjith PG, Choi SK (2013) Coal cleat permeability for gas movement under triaxial, non-zero lateral strain condition: a theoretical and experimental study. Fuel 109:389–399
4. Perera MSA (2017) Influences of CO_2 injection into deep coal seams: a review. Energy Fuels 31(10):10324–10334
5. Zhang XG, Ranjith PG, Perera MSA, Ranathunga AS, Haque A (2016) Gas transportation and enhanced coalbed methane recovery processes in deep coal seams: a review. Energy Fuels 30:8832–8849
6. Zhang Y, Lebedev M, Sarmadivaleh M, Barifcani A, Iglauer S (2016) Swelling-induced changes in coal microstructure due to supercritical CO_2 injection. Geophys Res Lett 43:9077–9083
7. Pini R, Ottiger S, Burlini L, Storti G, Mazzotti M (2009) Role of adsorption and swelling on the dynamics of gas injection in coal. J Geophys Res 114(B4):B04203
8. Heller R, Zoback M (2014) Adsorption of methane and carbon dioxide on gas shale and pure mineral samples. J Unconv Oil Gas Resour 8:14–24
9. Sang G, Elsworth D, Liu S, Harpalani S (2017) Characterization of swelling modulus and effective stress coefficient accommodating sorption-induced swelling in coal. Energy Fuels 31:8843–8851
10. Chen T, Feng XT, Pan Z (2018) Experimental study on kinetic swelling of organic-rich shale in CO_2, CH_4 and N_2. J Nat Gas Sci Eng 55:406–417
11. Day S, Fry R, Sakurovs R (2012) Swelling of coals in carbon dioxide, methane and their mixtures. Int J Coal Geol 93(1):40–48
12. Zarębska K, Ceglarska-Stefańska G (2008) The change in effective stress associated with swelling during carbon dioxide sequestration on natural gas recovery. Int J Coal Geol 74:167–174
13. Majewska Z, Ceglarska-Stefańska G, Majewski S, Ziętek J (2009) Binary gas sorption/desorption experiments on a bituminous coal: simultaneous measurements on sorption kinetics, volumetric strain and acoustic emission. Int J Coal Geol 77:90–102
14. Liu SM, Wang Y, Harpalani S (2016) Anisotropy characteristics of coal shrinkage/swelling and its impact on coal permeability evolution with CO_2 injection. Greenhouse Gases: Sci Technol 6:1–18
15. Niu Q, Cao L, Sang S, Zhou X, Wang Z, Wu Z (2017) The adsorption-swelling and permeability characteristics of natural and reconstituted anthracite coals. Energy 141:2206–2217
16. Niu Q, Cao L, Sang S, Zhou X, Wang Z (2018) Anisotropic adsorption swelling and permeability characteristics with injecting CO_2 in coal. Energy Fuels 32(2):1979–1991
17. Harpalani S, Chen G (1995) Estimation of changes in fracture porosity of coal with gas emission. Fuel 74(10):1491–1498
18. Levine JR (1996) Model study of the influence of matrix shrinkage on absolute permeability of coal bed reservoirs. Geol Soc Lond Spec Publ 109(1):197–212
19. George JD, Barakat MA (2001) The change in effective stress with shrinkage from gas desorption in coal. Int J Coal Geol 45(2–3):105–113

20. Connell LD, Lu M, Pan Z (2010) An analytical coal permeability model for tri-axial strain and stress conditions. Int J Coal Geol 84(2):103–114
21. Pan Z, Connell LD (2007) A theoretical model for gas adsorption-induced coal swelling. Int J Coal Geol 69(4):243–252
22. Zang J, Wang K (2017) Gas sorption-induced coal swelling kinetics and its effects on coal permeability evolution: model development and analysis. Fuel 189:164–177
23. Shi JQ, Durucan S (2004) Drawdown induced changes in permeability of coalbeds: a new interpretation of the reservoir response to primary recovery. Transp Porous Media 56:1–16
24. Cui X, Bustin AMM, Bustin RM (2009) Measurements of gas permeability and diffusivity of tight reservoir rocks: different approaches and their applications. Geofluids 9:208–223
25. Cui X, Bustin RM (2005) Volumetric strain associated with methane desorption and its impact on coalbed gas production from deep coal seams. AAPG Bull 89:1181–1202
26. Liu HH, Rutqvist J (2010) A new coal-permeability model: internal swelling stress and fracture-matrix interaction. Transp Porous Media 82(1):157–171
27. Wang S, Elsworth D, Liu J (2011) Permeability evolution in fractured coal: the roles of fracture geometry and water-content. Int J Coal Geol 87(1):13–25
28. Kumar H, Elsworth D, Mathews JP, Marone C (2016) Permeability evolution in sorbing media: analogies between organic-rich shale and coal. Geofluids 16:43–55
29. Fan J, Feng R, Wang J, Wang Y (2017) Laboratory investigation of coal deformation behavior and its influence on permeability evolution during methane displacement by CO_2. Rock Mech Rock Eng 50:1725–1737
30. Meng Y, Li Z (2017) Triaxial experiments on adsorption deformation and permeability of different sorbing gases in anthracite coal. J Nat Gas Sci Eng 46:59–70
31. Cao C, Li T, Shi J, Zhang L, Fu S, Wang B, Wang H (2016) A new approach for measuring the permeability of shale featuring adsorption and ultra-low permeability. J Nat Gas Sci Eng 30:548–556
32. Feng R, Harpalani S, Pandey R (2016) Evaluation of various pulse-decay laboratory permeability measurement techniques for highly stressed coals. Rock Mech Rock Eng 50:297–308
33. Feng R, Liu J, Harpalani S (2017) Optimized pressure pulse-decay method for laboratory estimation of gas permeability of sorptive reservoirs: part 1-background and numerical analysis. Fuel 191:555–564
34. Feng R, Harpalani S, Liu J (2017) Optimized pressure pulse-decay method for laboratory estimation of gas permeability of sorptive reservoirs: part 2-experimental study. Fuel 191:565–573
35. Anggara F, Sasaki K, Sugai Y (2016) The correlation between coal swelling and permeability during CO_2 sequestration: a case study using Kushiro low rank coals. Int J Coal Geol 166:62–70
36. Luffel DL, Guidry FK (1992) New core analysismethods for measuring reservoir rock properties of Devonian shale. J Petrol Technol 44(11):1184–1190
37. Suarez-Rivera R, Chertov M, Willberg D, Green S, Keller J (2012) Understanding permeability measurements in tight shales promotes enhanced determination of reservoir quality. Presented at the SPE Canadian unconventional resources conference, Calgary, Alberta, Canada, 30 Oct–1 Nov 2012
38. Goodman AL, Busch A, Duffy G, Fitzgerald JE, Gasem KAM, Gensterblum Y, Krooss BM, Levy J, Ozdemir E, Pan Z, Robinson JRL, Schroeder K, Sudibandriyo M, White C (2004) An inter-laboratory comparison of CO_2 isotherms measured on argonne premium coal samples. Energy Fuels 18:1175–1182
39. Zhang DF, Cui YJ, Liu B, Li SG, Song WL, Lin WG (2011) Supercritical pure methane and CO_2 adsorption on various rank coals of China: experiments and modeling. Energy Fuels 25:1891–1899
40. Huo P, Zhang D, Yang Z, Li W, Zhang J, Jia S (2017) CO2 geological sequestration: displacement behavior of shale gas methane by carbon dioxide injection. Int J Greenhouse Gas Control 66:48–59
41. Sakurovs R, Day S, Weir S, Duffy G (2007) Application of a modified Dubinin-Radushkevich equation to adsorption of gases by coals under supercritical conditions. Energy Fuels 21:992–997

42. Jia J, Sang S, Cao L, Liu S (2018) Characteristics of CO_2/supercritical CO_2 adsorption-induced swelling to anthracite: an experimental study. Fuel 216:639–647
43. Yang Z, Sang Q, Dong M, Zhang S, Li Y, Gong H (2015) A modified pressure-pulse decay method for determining permeabilities of tight reservoir cores. J Nat Gas Sci Eng 27:236–246
44. Brace WF, Walsh JB, Frangos WT (1968) Permeability of granite under high pressure. J Geophys Res 73:2225–2236
45. Cao P, Liu J, Leong YK (2017) A multiscale-multiphase simulation model for the evaluation of shale gas recovery coupled the effect of water flowback. Fuel 199:191–205
46. Liu J, Wang JG, Gao F, Ju Y, Zhang X, Zhang L (2016) Flow consistency between non-Darcy flow in fracture network and nonlinear diffusion in matrix to gas production rate in fractured shale gas reservoirs. Transp Porous Media 111:97–121
47. Ma T, Rutqvist J, Oldenburg CM, Liu W, Chen J (2017) Fully coupled two-phase flow and poromechanics modeling of coalbed methane recovery: impact of geomechanics on production rate. J Nat Gas Sci Eng 45:474–486
48. Yang Z, Dong M, Zhang S, Gong H, Li Y, Long F (2016) A method for determining transverse permeability of tight reservoir cores by radial pressure pulse decay measurement. J Geophys Res Solid Earth 121:7054–7070
49. Harpalani S, Chen G (1997) Influence of gas production induced volumetric strain on permeability of coal. Geotech Geol Eng 15:303–325
50. Liu SM, Harpalani S (2014) Compressibility of sorptive porous media: part 2 experimental study on coal. AAPG Bull 98:1773–1788
51. Lin J, Ren T, Wang G, Booth P, Nemcik J (2017) Experimental study of the adsorption-induced coal matrix swelling and its impact on ECBM. J Earth Sci 28:917–925
52. Chen T, Feng XT, Pan Z (2015) Experimental study of swelling of organic rich shale in methane. Int J Coal Geol 150:64–73
53. Guo X, Wang ZM, Zhao YL (2016) A comprehensive model for the prediction of coal swelling induced by methane and carbon dioxide adsorption. J Nat Gas Sci Eng 36:563–572
54. Day S, Fry R, Sakurovs R (2008) Swelling of Australian coals in supercritical CO_2. Int J Coal Geol 74:41–52
55. Siriwardane H, Haljasmaa I, McLendon R, Irdi G, Soong Y, Bromhal G (2009) Influence of carbon dioxide on coal permeability determined by pressure transient methods. Int J Coal Geol 77(1):109–118
56. Sakhaee-Pour A, Bryant S (2012) Gas permeability of shale. SPE Reserv Eval Eng 15:401–409
57. Wang Y, Liu S, Elsworth D (2015) Laboratory investigations of gas flow behaviors in tight anthracite and evaluation of different pulse-decay methods on permeability estimation. Int J Coal Geol 149:118–128
58. Zhang Y, Lebedev M, Sarmadivaleh M, Barifcani A, Rahman T, Iglauer S (2016) Swelling effect on coal micro structure and associated permeability reduction. Fuel 182:568–576

BY NC ND

Chapter 5
A Large Pressure Pulse Decay Method to Simultaneously Measure Permeability and Compressibility of Tight Rocks

5.1 Introduction

Permeability characteristics of oil and gas strata have become one of the key points for its exploitation and transportation [1–4]. Due to the extremely low permeability of tight rock, it is very time-consuming to reach a steady state when using the steady-state method for permeability measurement, making it difficult to be applied in engineering [5–9]. In order to overcome this disadvantage, Brace et al. [10] put forward the pressure pulse transient decay method to measure the permeability characteristics of tight rocks for the first time. Compared with the steady-state method, the transient method can effectively shorten the permeability testing time and obtain relatively reliable results. However, two important assumptions were made in Brace's method: (1) the pore volume of rock is ignored; (2) the pressure gradient varies with time but remains unchangeable along the rock sample's length. Hsieh et al. [11] derived the exact solution of Brace's model without the two assumptions. However, the obtained solution is too complex to be used for calculation in engineering. Dicker and Smits [12]optimized the calculation conditions of the transient pulse method and obtained an approximate solution that is easier for engineering application. To further simplify Dicker and Smits's solution, Jones [13]optimized the computation by introducing a factor "*f*" and correcting the role of compressive storage to make the form of the solution the same as Brace's solution. Recently, researchers have proposed some modified pressure decay methods, such as the one chamber pressure-pulse decay method [14], bi-directional pressure-pulse decay method [15], and transverse pressure-pulse decay method [16, 17]. Table 5.1 lists a summary of the pressure pulse decay method, which is relevant to this study.

Note that all these methods in Table 5.1 are applicable to the case of small initial pulse pressure (no more than 10% of initial reservoir pressure recommended by Brace et al. 1968) due to the limitation of the following two aspects [10]: first, for transient tests with large pulse pressure, effective stress decreases considerably (even though confining pressure keeps constant), resulting in a change of rock permeability,

C. Wang and Y. Zhao, *Permeability Measurement of Tight Rock*,
https://doi.org/10.1007/978-981-92-0597-4_5

Table 5.1 Development of the pressure pulse-decay method for permeability measurement

Author	Year	Novelties	Comments
Brace et al.	1968	First proposed pressure pulse-decay method	The pore volume of rock is ignored
Hsieh et al.	1981	Obtained the exact solution of Brace's model	Considered compressive storage by using water
Dicker and Smits	1988	Obtained an approximate solution by optimizing Hsieh et al.'s solution	Applied compressive storage effect into permeability calculation
Cui et al.	2009	Demonstrated the effect of adsorption on the measurements of permeability	Considered both compressive storage and adsorption
Yang et al.	2015	Proposed the one chamber pressure-pulse decay method	Simpler experimental apparatus, and easier operation
Hannon	2016	Developed bi-directional pressure-pulse decay method	Faster without sacrificing on the accuracy and reliability of parameter estimates
Yang et al.	2016a	Proposed transverse pressure pulse decay method	Determining transverse permeability of tight rocks
Wang et al.	2020	Simultaneously determined permeability, porosity, and adsorption capacity	Considered both compressive storage and adsorption

that is, permeability is not a constant for large pressure pulse decay tests; second, the compressibility and viscosity of the fluid, especially for gas, changes with the increase of the pore pressure. These two factors limit the transient pulse decay method to choose a small pressure difference. However, a small pressure difference also means that it will take a longer time to obtain the later-time pressure attenuation curve that can be used to calculate permeability (Brace's method of calculating permeability is not suitable for the early-time pressure attenuation curve).

For porous mediums, the permeability varies with effective stress, and the relation between them is generally accepted to be expressed in exponential form [18]. Hence, the relationship between permeability and effective stress can be described as:

$$k = k_0 e^{-3c_p(\sigma - \sigma_0)} \tag{5.1}$$

where c_p is the pore volume compressibility; σ and σ_0 are the effective stress and initial effective stress, respectively; k_0 is the initial permeability at effective stress σ_0. Equation (5.1) includes two important parameters (k_0, c_p) with clear physical meaning that are very important for engineering.

The pore volume compressibility (c_p) of rock is an important parameter in the analysis of oil and gas fields [19–22], which is usually measured by volume method. However, this method is time-consuming. Hall [23]presented the relationship between the initial porosity and the pore volume compressibility, and drew it into a curve to facilitate engineering applications, known as the "Hall plot". However, the method of measuring pore volume compressibility through the volume method faces two following challenges: first, pore volume compressibility decreases with the increase of rock porosity; second, the measured pore volume compressibility is usually greater than that of the reservoir fluid. Recently, some new methods for measuring pore volume compressibility were proposed. For example, Zhu et al. [24] deduced a new analytical model to obtain the pore volume compressibility of porous rock; Sui et al. [25] used the spherical pore model of digital rock to estimate the pore volume compressibility.

Inspired by previous studies, this work presented a large pressure pulse decay method (LPPD) to simultaneously estimate permeability and pore volume compressibility. In this method, the pressure pulse is much larger than 10% of the initial reservoir pressure. Water and gas are commonly used as the testing fluids in the permeability tests. But the gas compressibility varies with pressure, and adsorption effects and Klinkenberg effects can also affect the measured permeability. Although the LPPD method is applicable for both compressible/incompressible fluids, water is used as the testing fluid in the permeability tests for simplicity. In this study, a series of transient tests with large pressure pulse were conducted on a coal sample through water. The proposed method was applied to process experimental data obtained from large pressure decay tests under different confining pressures. Then, the estimated permeability and pore volume compressibility were compared and verified with Brace's method and experimental results obtained from measured deformation parameters, respectively.

5.2 Mathematical Model for the Large Pressure Pulse Decay Test

5.2.1 Mathematical Model

In a pressure pulse decay test, the pressure of the test system is first kept at an initial pressure P_0. After that, the pressure of the upstream chamber is significantly increased to create a large pressure difference, which made the fluid to flow from upper boundary to lower boundary through microcracks and pore throats of the sample. Assuming that the fluid flow through the core samples subjected to large pressure pulse obeys Darcy's law [26], fluid flow through a microelement (with length Δx and cross-sectional area A, shown in Fig. 5.1) can be described as

$$q = -\frac{k}{\mu}\frac{\partial P}{\partial x}A \tag{5.2}$$

where q and μ represent the flow rate and the fluid viscosity, respectively; k represents the permeability of the sample and P is the pore pressure.

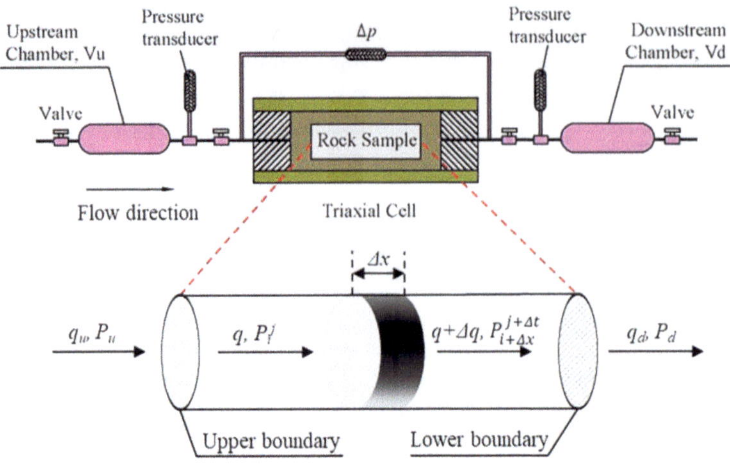

Fig. 5.1 Schematic diagram of rock sample for fluid permeability experimental modeling

Since the confining pressure keeps constant during each transient test, the change of effective stress is related to the pore pressure evolution. Therefore, Eq. (5.1) can be replaced by the pore pressure based on Biot's effective stress law (Note that Biot's coefficient is considered to be one in this study following previous studies [27–29]

$$k = k_0 e^{3c_p(P-P_0)} \tag{5.3}$$

where P_0 is the initial pore pressure. Combining Eqs. (5.2) with (5.3), we have

$$q = -\frac{k_0 e^{3c_p(P-P_0)}}{\mu}\frac{\partial P}{\partial x}A \tag{5.4}$$

The flow increment Δq across the microelement of length Δx during a time increment Δt can be calculated:

$$dqdt = -\frac{k_0 A}{\mu}e^{3c_p(P-P_0)}\left[\frac{\partial^2 P}{\partial x^2} + 3c_p\left(\frac{\partial P}{\partial x}\right)^2\right]dtdx \tag{5.5}$$

Due to the compressibility of the fluid and the solid, there will be two parts in the net storage of fluid in the microelement of length Δx during a time increment Δt. The total storage during a time increment Δt can be given as follows

$$\frac{\partial\left(\frac{V_p}{L}dx\right)}{\partial t}dt + \left(\frac{V_p}{L}dx\right)\beta\frac{\partial P}{\partial t}dt \tag{5.6}$$

where V_{p} represents the whole pore volume of the sample with length L, β represents the fluid compressibility (which can be regarded as a constant if water is used as the testing fluid). The first part in Eq. (5.6) is equal to the storage of fluid due to compression of the solid matrix and the second part represents storage due to compression of the fluid [10].

In order to maintain fluid continuity during flowing, the net flow increment $-dqdt$ given by Eq. (5.5) must equal the net storage shown by Eq. (5.6)

$$\left(\frac{\partial \varphi V_b}{V_b \partial P} + \varphi\beta\right)\frac{\partial P}{\partial t} = \frac{k_0}{\mu}e^{3c_p(P-P_0)}\left[\frac{\partial^2 P}{\partial x^2} + 3c_p\left(\frac{\partial P}{\partial x}\right)^2\right] \tag{5.7}$$

where V_b represents the bulk volume, $V_p = \phi V_b$, ϕ represents the porosity of the sample. Both the bulk volume and the porosity change are related to pore pressure [30]

$$\frac{\partial \varphi V_b}{V_b \partial P} = \frac{\partial \varphi}{\partial P} + \varphi c_b \tag{5.8}$$

where c_b represents the bulk compressibility of the sample. The bulk volume of sample contains two parts: the pore volume V_p and the solid volume V_s

$$V_{\mathrm{b}} = V_p + V_s \tag{5.9}$$

Differentiating Eq. (5.9) with pore pressure P

$$c_b = \varphi c_p + (1-\varphi)c_s \tag{5.10}$$

where c_s is solid matrix compressibility, which is much smaller than bulk compressibility c_b and pore volume compressibility c_p. For simplicity, Eq. (5.10) can be rewritten as [31]

$$c_b = \varphi c_p \tag{5.11}$$

The rock pore volume compressibility is defined as

$$c_p = \frac{\partial V_p}{V_p \partial P} = \frac{\partial \varphi V_b}{\varphi V_b \partial P} = \frac{V_b \partial \varphi + \varphi \partial V_b}{\varphi V_b \partial P} = \frac{\partial \varphi}{\varphi \partial P} + c_b \tag{5.12}$$

Substituting the Eqs. (5.11) into (5.12), yields

$$\frac{\partial \varphi}{\partial P} = \varphi(1-\varphi)c_p \tag{5.13}$$

Combining the Eqs. (5.7), (5.8), (5.10), and (5.13), we have

$$\varphi\left(c_p+\beta\right)\frac{\partial P}{\partial t}=\frac{k_0}{\mu}e^{3c_p(P-P_0)}\left[\frac{\partial^2 P}{\partial x^2}+3c_p\left(\frac{\partial P}{\partial x}\right)^2\right] \tag{5.14}$$

For most tight rocks, the pore volume compressibility c_p is very small, hence compared to $\frac{\partial^2 P}{\partial x^2}$ in Eq. (5.14), $c_p\left(\frac{\partial P}{\partial x}\right)^2$ is higher-order small and can be negligible. Therefore Eq. (5.14) can be changed to

$$\frac{\partial P}{\partial t}=\frac{k_0}{\varphi\mu\left(\beta+c_p\right)}e^{3c_p(P-P_0)}\frac{\partial^2 P}{\partial x^2} \tag{5.15}$$

For the upper boundary, the quantity of fluid flowing out of the upstream chamber is equal to the amount of fluid passing the upper boundary into the sample, which can be expressed by the following equation

$$\frac{dP_u}{dt}=\frac{k_0}{\mu\beta}\frac{A}{V_u}e^{3c_p(P-P_0)}\left.\frac{\partial P}{\partial x}\right|_{x=0} \tag{5.16}$$

where P_u and V_u are the pressure and volume of the upstream chamber, respectively.

For the lower boundary, in the same way, the increase in the quantity of fluid in the downstream chamber is the same as the amount of fluid flowing out of the surface of the sample that can be expressed by the following equation

$$\frac{dP_d}{dt}=-\frac{k_0}{\mu\beta}\frac{A}{V_u}e^{3c_p(P-P_0)}\left.\frac{\partial P}{\partial x}\right|_{x=L} \tag{5.17}$$

where P_d and V_d are the pressure and volume of the downstream chamber, respectively. Considering the initial fluid pressure of upstream and downstream chambers at the beginning of the test, it can be given as

$$\begin{aligned}P_u(t_0)&=P_0+\Delta P\\P(x,t_0)&=P_0\end{aligned} \tag{5.18}$$

where $P_u(t_0)$ and $P(x,t_0)$ are the pressure of the upstream chamber and the pore pressure of the rock sample at the initial time t_0, respectively, P_0 is the initial equilibrium pore pressure at time t_0, ΔP is the instantaneous increment pressure of the upstream chamber at time t_0.

The equation set (including the controlling equation Eq. (5.15), boundary conditions Eqs. (5.16) and (5.17), and initial conditions Eq. (5.18)) is a mathematical model for investigating fluid flow behavior in a rock sample. The controlling equation is a partial differential equation with variable coefficients and complicated boundary conditions, making the equation set very difficult to get analytic solutions. Therefore, it is necessary and practical to introduce the finite difference method to get the numerical solution of the equation set.

5.2.2 Methodology of Numerical Solution

Discretization of the continuous mathematical model mentioned above is required to solve partial differential equations through the finite difference method. The variation of pore pressure in the sample can be divided into two dimensions space: time and length. On a two-dimensional plane, the X-axis in the horizontal direction and T-axis in the vertical direction are established to represent the length of the sample and the time of the test, the two axes are orthogonal to each other. Both the length space [0 L] on the X-axis and the time–space [0 T] on the T-axis are respectively divided into equal segments by $i = 0, 1, 2, \ldots, I$ and $j = 0, 1, 2, \ldots, J$, the length of the segment in the X-axis is Δx and Δt in the T-axis. Therefore, as shown in Fig. 5.2, taking P_i^j as an example, it represents the pore pressure at the point of length $i\Delta x$ and time $j\Delta t$.

Using the finite difference method, the controlling equation Eq. (5.15) can be expressed differentially as

$$\frac{P_i^{j+1} - P_i^j}{\Delta t} = \frac{k_0}{\varphi\mu\left(\beta + c_p\right)} \frac{e^{3c_p\left(P_{i+1}^j - P_0\right)} P_{i+1}^j - 2e^{3c_p\left(P_i^j - P_0\right)} P_i^j + e^{3c_p\left(P_{i-1}^j - P_0\right)} P_{i-1}^j}{\Delta x^2} \tag{5.19}$$

The two boundary conditions Eqs. (5.16), (5.17) can be replaced by

$$\frac{P_1^{j+1} - P_1^j}{\Delta t} = \frac{k_0}{\mu\beta} \frac{A}{V_u} \frac{e^{3c_p(P_2^j - P_0)} P_2^j - e^{3c_p(P_1^j - P_0)} P_1^j}{\Delta x} \tag{5.20}$$

$$\frac{P_I^{j+1} - P_I^j}{\Delta t} = -\frac{k_0}{\mu\beta} \frac{A}{V_d} \frac{e^{3c_p(P_{I-1}^j - P_0)} P_{I-1}^j - e^{3c_p(P_I^j - P_0)} P_I^j}{\Delta x} \tag{5.21}$$

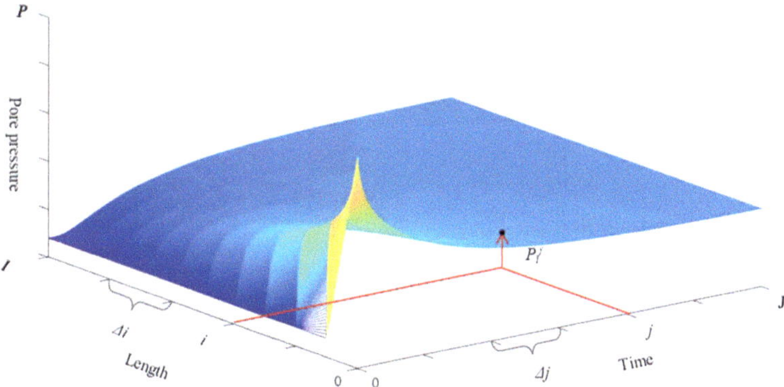

Fig. 5.2 Schematic diagram for principle of numerical simulation

The above finite difference equations and initial conditions (Eqs. 5.19–5.21) can be written in MATLAB language to get the numerical solution. However, it is worth noting that these equations contain two unknown parameters (the initial permeability of the sample k_0 and the pore volume compressibility of the sample c_p).

5.2.3 *Unknown Parameters Estimation and Global Convergence*

The fitting of unknown parameters can be obtained by using the least-squares method which can be expressed as [32]:

$$\min f\left[P\left(x, t_j\right)\right] = \frac{1}{2}\sum_{j=1}^{n} r^2\left[P\left(x, t_j\right)\right] \tag{5.22}$$

where $r[P(x, t_j)] = P_E(x, t_j) - P_M(x, t_j)$ is the residual, $P_E\left(x, t_j\right)$ is the pore pressure measured by experiment at time t_j and length x, and $P_M\left(x, t_j\right)$ is the pore pressure of numerical solution calculated by MATLAB language mentioned above at time t_j and length x, which contain the unknown parameters (k_0, c_p); n represents the number of time points in the experiment; and $f\left[P\left(x, t_j\right)\right]$ is the sum of squared residuals. Therefore, acquiring the two unknown parameters is essentially a mathematical problem of obtaining the optimal solution. The least-square method applied to this parameters estimation problem can be given as follows

$$\min f\left[P\left(x, t_j\right)\right] = \frac{1}{2}\sum_{j=1}^{n}\left[P_E\left(x, t_j\right) - P_M\left(x, t_j\right)\right]^2 \tag{5.23}$$

For the sake of finding the minimum of the function $f\left[P\left(x, t_j\right)\right]$, two prerequisites must be met. One prerequisite is that the derivatives of $f\left[P\left(x, t_j\right)\right]$ with respect to k_0 and c_p should be zero:

$$\nabla f\left[P\left(x, t_j\right)\right] = J\left[P\left(x, t_j\right)\right]^T\left[P_E\left(x, t_j\right) - P_M\left(x, t_j\right)\right] = 0 \tag{5.24}$$

where $J\left[P\left(x, t_j\right)\right] = \left\{\frac{\partial\left[P_E\left(x,t_j\right) - P_M\left(x,t_j\right)\right]}{\partial k_0}, \frac{\partial\left[P_E\left(x,t_j\right) - P_M\left(x,t_j\right)\right]}{\partial c_p}\right\}$ is the Jacobian matrix. Another prerequisite is that the Hessian matrix for $f[P(x, t_i)]$:

$$\nabla^2 f\left[P\left(x, t_j\right)\right] = J\left[P\left(x, t_j\right)\right]^T J\left[P\left(x, t_j\right)\right] + \sum_{j=1}^{n} r\left[P\left(x, t_j\right)\right]\nabla^2 r\left[P\left(x, t_j\right)\right] \tag{5.25}$$

Equations (5.24) and (5.25) are used to ensure that $f\left[P\left(x,t_j\right)\right]$ is strictly a locally optimal solution of $P^*(x,t_j)$. There are many algorithms for solving nonlinear least-square questions. Specific to this work, Levenberg–Marquardt (L-M) method is adopted, and its iterative method can be expressed by the following equation:

$$P_{z+1}(x,t_j) = P_z(x,t_j) - \left\{\nabla^2 f[P_z(x,t_j)] + \mu_z I\right\}^{-1} \nabla f[P_z(x,t_j)] \tag{5.26}$$

where the subscript z represents the number of steps in the iteration, μ_z is a scalar parameter which decreases gradually as the iteration process is closed to the local minimum of $P^*(x,t_j)$, I is an identity matrix. For the convenience of calculation, the second term in Eq. (5.25) can be ignored, because $r[P(x,t_j)]$ is relatively small in least-squares problems. So Eq. (5.26) can be changed into:

$$\begin{aligned} P_{z+1}(x,t_j) &= P_z(x,t_j) \\ &\quad - \left\{J^T[P_z(x,t_j)]J[P_z(x,t_j)] + \mu_z I\right\}^{-1} J^T[P_z(x,t_j)][P_E(x,t_j) - P_M(x,t_j)] \end{aligned} \tag{5.27}$$

However, the L-M method is not sufficient to get the optimal solution of the problem because Eq. (5.27) is not a monotonically increasing or decreasing function, which means that it may have many local minimums solutions and that the method of local convergence will not obtain the optimal solution of global convergence [33, 34]. Therefore, a global convergence optimization method within a reasonable range of unknown parameters should be used to solve this problem. The following procedure and a flow chart (Fig. 5.3) are established to calculate the global optimal solution.

Step 1 Selecting iteration parameters s, $g \in (0,1)$, the two initial iteration values of k_0 and c_p which should be within the reasonable range of the sample, allowable error δ. In our procedure, s, g, and δ are 0.8, 0.4, and 1×10^{-6}, respectively. The first step will start at $z = 0$.

Step 2 Calculating $\nabla f[P_z(x,t_j)]$ in Eq. (5.24). If the result is less than δ, then stop the iterative calculation and output the values of the two estimation parameters k_0 and c_p as the approximation solution.

Step 3 Solving the following equation

$$\left\{J^T\left[P_z\left(x,t_j\right)\right]J\left[P_z\left(x,t_j\right)\right] + \mu_z I\right\}d_z = -J^T\left[P_z\left(x,t_j\right)\right]\left[P_E\left(x,t_j\right) - P_M\left(x,t_j\right)\right], \tag{5.28}$$

and get d_z.

Step 4 Let m_z be the smallest non-negative integer m satisfying the following inequality.

$$f[P_z(x,t_j) + g^m d_z] \le f[P_z(x,t_j)] + s g^m \nabla f[P_z(x,t_j)]^T d_z, \tag{5.29}$$

and let $s_z = g^{m_z}$, $P_{z+1}(x,t_j) = P_z(x,t_j) + s_z d_z$.

Step 5 Let $\mu_z = \|P_E - P_{zM}\|$, $z = z + 1$, return to *step 2*.

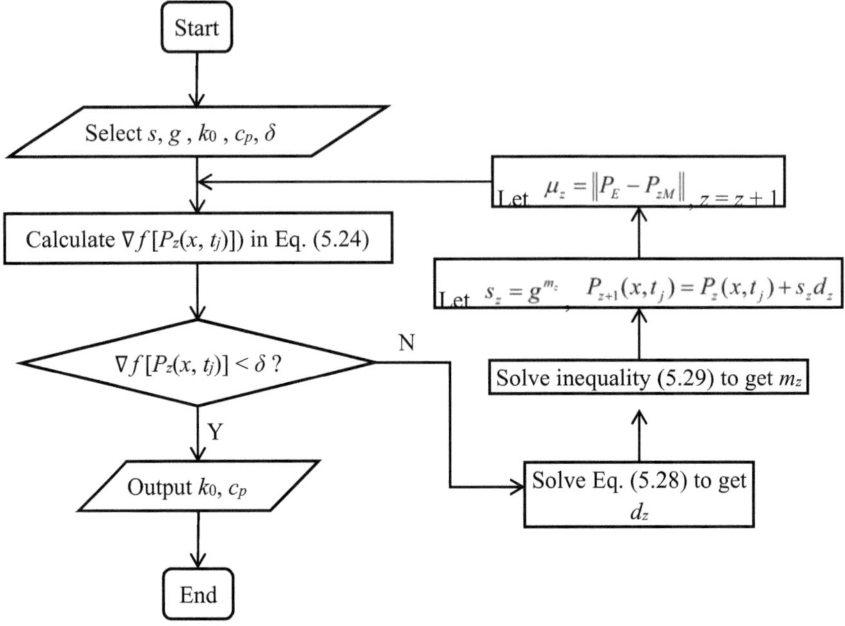

Fig. 5.3 The flow chart of calculating the global optimal solution

Based on the above procedure, we can get the optimal solution of global convergence to fit the experimental data. The smaller the allowable error δ, the more accurate the calculation will be. But the disadvantage is that as the allowable error decreases, the calculation will become more time-consuming. After many times of iterations and analyses, a good result can be obtained when the allowable error is set at 1×10^{-6}.

For simplicity, the fluid pressure difference between up-/downstream chambers can be used as the fitting data. Thus $P_E(x, t_j)$ in Eq. (5.23) can be expressed as $\Delta P_E(t_j) = P_{Eu}(t_j) - P_{Ed}(t_j)$, where $P_{Eu}(t_j)$ and $P_{Ed}(t_j)$ are the upstream ($x = 0$) and downstream ($x = L$) chambers pressure, respectively. Similarly, $P_M(x, t)$ in Eq. (5.23) can be expressed as $\Delta P_M(t_j) = P_{Mu}(t_j) - P_{Md}(t_j)$, where $P_{Mu}(t_j)$ and $P_{Md}(t_j)$ are the numerical solution of upstream and downstream chamber pressure, respectively. Through the process shown in the flow chart (Fig. 5.3), two unknown parameters (k_0, c_p) can be fitted. Furthermore, by substituting the two parameters into the Eqs. (5.19)–(5.21), the fluid pressure of the sample at any point and any time during the test can be calculated.

5.2.4 Modified Brace's Method

Brace et al. [10]first presented the transient pulse decay method for estimating rock permeability, and the simplified solution can be given by the following equations

$$P_u(t) - P_d(t) = \Delta P(t_0)e^{-\alpha t} \tag{5.30}$$

$$\alpha = \frac{kA}{\mu\beta L}\left(\frac{1}{V_u} + \frac{1}{V_d}\right) \tag{5.31}$$

where α is the slope of the line with time as the horizontal axis and $(P_u(t) - P_d(t))/\Delta P$ as the vertical axis on 2D semi-log plane. Based on Brace's model, Dicker and Smits [12] developed a general solution by considering the compressive storage of fluid in the sample.

During the large pulse pressure decay testing process, both fluid and pore volume compressibility should be considered. Therefore, for fluid whose compressibility varies little with pressure Brace's method is modified by adding the pore volume compressibility to the controlling equation:

$$\frac{\partial P(x,t)}{\partial t} = \frac{k}{\varphi\mu(\beta + c_p)}\frac{\partial^2 P(x,t)}{\partial x^2} \tag{5.32}$$

$$\frac{dP_u}{dt} = \frac{k}{\varphi\mu\beta L}\frac{V_p}{V_u}\left.\frac{\partial P(x,t)}{\partial x}\right|_{x=0} \tag{5.33}$$

$$\frac{dP_d}{dt} = -\frac{k}{\varphi\mu\beta L}\frac{V_p}{V_u}\left.\frac{\partial P(x,t)}{\partial x}\right|_{x=L} \tag{5.34}$$

The initial conditions and boundary conditions of the above modified mathematical model Eqs. (5.33)–(5.34) are the same as Brace's model. By replacing the fluid compressibility with the sum of fluid and pore volume compressibility, three dimensionless parameters a, b, and t_D are defined as

$$a = \frac{(\beta + c_p)V_p}{\beta V_u},\ b = \frac{(\beta + c_p)V_p}{\beta V_d},\ t_D = \frac{kt}{\varphi(\beta + c_p)\mu L^2} \tag{5.35}$$

where a and b are ratios of the pore volume compressive storage of the sample to the upstream and downstream volumes compressive storage, respectively, and t_D is dimensionless time. Thus, the general solution developed by Dicker and Smits [12] can be changed into

$$\Delta P_D(a, b, t_D) = 2\sum_{m=1}^{\infty} \exp\left(-t_D\theta_m^2\right) \cdot \frac{a\left(b^2+\theta_m^2\right) - (-1)^m b\sqrt{\left(a^2+\theta_m^2\right)\left(b^2+\theta_m^2\right)}}{\theta_m^4 + \theta_m^2\left(a + a^2 + b + b^2\right) + ab(a + b + ab)} \tag{5.36}$$

$$\tan\theta = \frac{(a+b)\theta}{\theta^2 - ab} \tag{5.37}$$

However, Eq. (5.36) is an evaluation of infinite series, and each series contains the roots of Eq. (5.37). Meanwhile, the roots of Eq. (5.37) depend on the tangent function of θ in a relatively unintuitive form [13]. To simplify the equations and make relationships among variables more intuitive, Jones [13] defined a factor "f"

$$f = \frac{\theta_1^2}{a+b} \tag{5.38}$$

$$\alpha = \frac{fkA}{\mu\beta L}\left(\frac{1}{V_u} + \frac{1}{V_d}\right) \tag{5.39}$$

where θ_1 represents the first root of the Eq. (5.37). The permeability obtained by Eq. (5.39) can be regarded as a modified Brace's method that considers the pore volume compressibility of the sample.

5.3 Experiment Methodology

5.3.1 Sample and Experimental Apparatus

A cylindrical coal sample with a diameter of 50 mm and a length of 100 mm was selected for the permeability test. The sample was taken from Hou Wenjialiang Coal Mine in Inner Mongolia, China. The large transient pulse decay tests were carried out on the MTS815 testing system which composed of three automatic servo devices: confining pressure, axial pressure, and pore pressure loading units. The system, shown in Fig. 5.4, was configured with a data collection unit that can automatically monitor stress, strain, and pore pressure during the test. The system was configured with two chambers (upstream chamber and downstream chamber) which can be used for transient pulse tests for estimating rock permeability. Volume values of the upstream and downstream chambers were 0.176 and 0.205 L, respectively.

When gas is used as the testing fluid, its physical properties vary with fluid pressure, and the equations of its properties changing with pressure can be substituted into the LPPD model. However, the complicated thing is that when gas flows through the sample, the non-negligible slippage and adsorption effects (adsorption capacity and model are difficult to determine accurately) can be accompanied, which will affect the accuracy of obtaining parameters. To simplify the process of verifying the

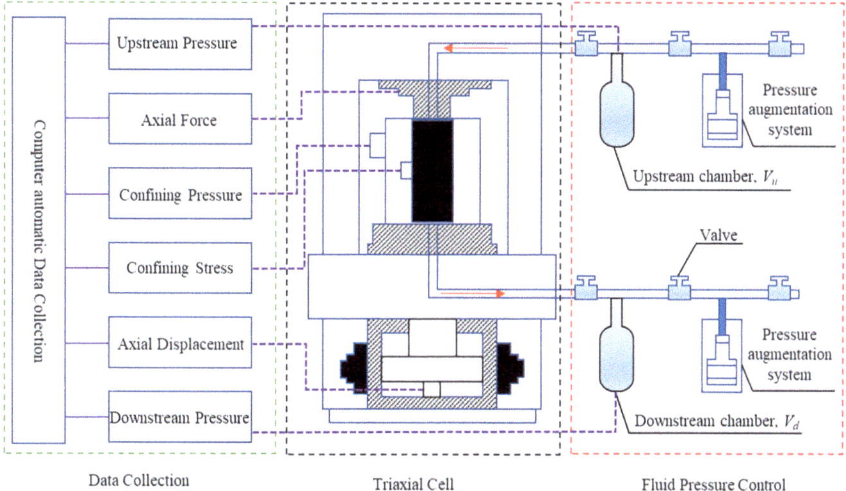

Fig. 5.4 Schematic diagram of experimental apparatus for rock permeability measurements

LPPD method, water was selected as the testing fluid due to the low compressibility and stable physical parameters under varied pressure conditions.

5.3.2 *Experimental Method and Procedures*

To verify the feasibility of the LPPD method and compare it with Brace's method, the unidirectional pressure pulse decay method was adopted in this test. Note that the LPPD method is also suitable for other pressure pulse decay methods, such as the bi-directional (or the transverse) pressure-pulse decay method mentioned in the introduction, as long as the boundary and initial conditions are changed correspondingly.

In order to verify the situation of large pressure difference proposed by the LPPD method, the initial fluid pressure was set as 0.2 MPa, and the instantaneous pressure increment in the upstream chamber at the initial time was 2 MPa with the pressure increased by 10 times. The instantaneous pressure increment is much higher than the restriction pressure condition of less than 10% in Brace's method. Eight groups of different confining pressures were set to verify the adaptability of the LPPD method under different effective stresses. To make the two parameters (k_0, c_p) comparable, those 8 groups of tests were all carried out on the same sample. The transient tests with large pressure pulse were carried out according to the following steps.

Step 1 Put the sample into the triaxial cell, and then load the confining pressure and axial pressure to the target value (at the same rate). The confining pressure ranged from 4 to 18 MPa in a gradual level of 2 MPa (i.e., 8 groups in total).

Step 2 Applied the initial water pressure in the up-/downstream chambers. The initial water pressure in the up-/downstream chambers was set to be 0.2 MPa.

Step 3 As the water pressure of the testing system became stable, the water pressure of the upstream chamber was suddenly increased to 2.2 MPa to provide an initial large pressure difference (2 MPa).

Step 4 Due to the pressure difference, the fluid in the upstream chamber would flow through the sample into the downstream chamber. The pressure difference decreased over time (until equilibrium was attained) and was automatically recorded by the testing system.

Step 5 Repeated steps (1) to (4) until completing all transient pulse tests.

5.4 Results and Discussions

5.4.1 Data Interpretations and Parameters Estimation

The experimental data were used to estimate initial permeability k_0 and pore volume compressibility c_p using the LPPD method. Figure 5.5 shows the experimental data and the corresponding fitting curve using the proposed globally convergent least square method. The dashed lines represent the calculated curves of water pressure in the up-/downstream chamber [$\Delta P_{Mu}(t)$and $\Delta P_{Md}(t)$]. The figures indicate that the calculation results from the numerical method of the LPPD method (Eqs. 5.19–21) match the experimental data quite well. For the 8 groups of experiments data (with varied confining pressure), the fitting coefficients are all larger than 0.99.

Table 5.2 lists the estimated values of k_0 and c_p. The initial permeability k_0 decreases as the confining pressure increases under the same test condition. This result can be explained by the fact that with the increase of confining pressure, the channel for fluid flow inside the sample is gradually narrowed or closed. Although these values show some discreteness for the estimated pore volume compressibility c_p, the deviation is small. The average value of the 8 tests groups is 3.273×10^{-8} Pa^{-1} and the standard deviation is 1.25×10^{-9}, which shows that the LPPD method can well acquire a relatively stable range of values of c_p.

5.4.2 Comparison of Estimated Permeability

Due to the large pulse pressure in the transient test, Eq. (5.1) reveals that the permeability should be relevant to the pore pressure. However, both Brace's method and the modified Brace's method only estimated a constant permeability in one fixed confining pressure test. To compare the three methods, an average permeability is defined as $\overline{k} = \left\{k_0 + k_0 e^{-3c_p[P_u(t_0)-P_0]}\right\}/2$ in the LPPD method. Figure 5.6 shows the estimated permeability using the LPPD method, Brace's method, and modified

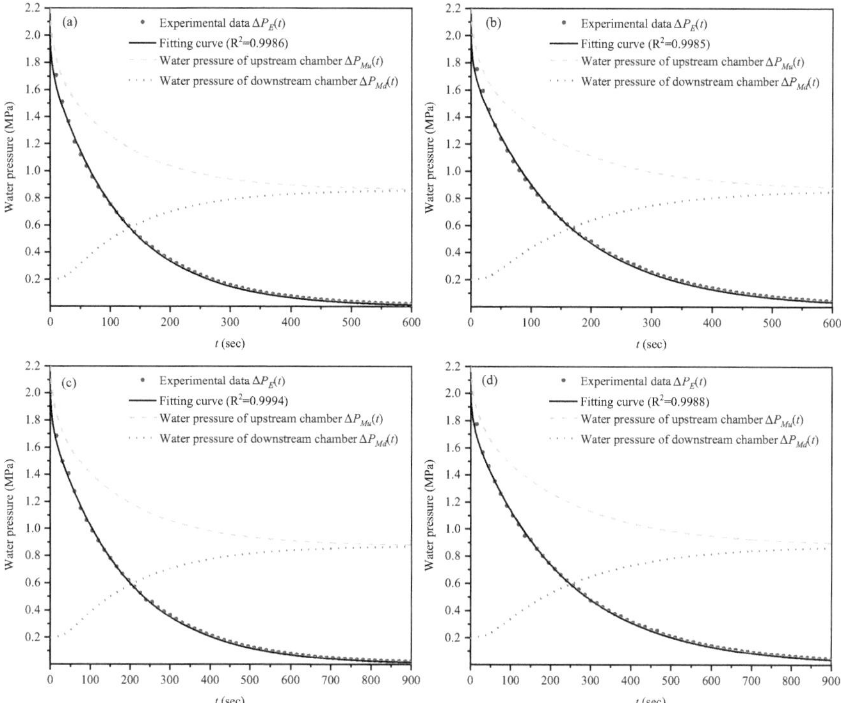

Fig. 5.5 Experimental data and fitting curve under different confining pressures **a** confining pressure 4 MPa; **b** confining pressure 6 MPa; **c** confining pressure 8 MPa; **d** confining pressure 10 MPa; **e** confining pressure 12 MPa; **f** confining pressure 14 MPa; **g** confining pressure 16 MPa; and **h** confining pressure 18 MPa

Brace's method. In Fig. 5.6, k_B is the estimated permeability using Brace's method, and k_M is the estimated permeability using the modified Brace's method. In addition, permeabilities calculated from the LPPD method are between Brace's method and the modified Brace's method. Since the pore volume and the pore volume compressibility of the sample are not considered in Brace's method, these assumptions result in the calculated permeabilities being smaller than the other two methods. Although pore volume compressibility is simply implanted into the modified Brace method, the permeability is assumed to be a constant in the modified Brace method, which also induces the discrepancy with the LPPD method.

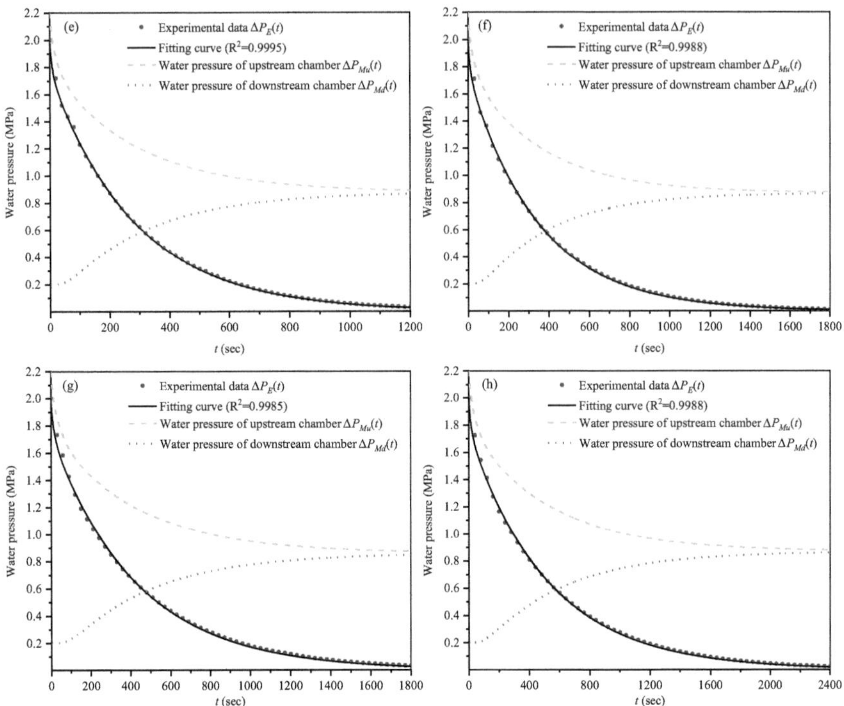

Fig. 5.5 (continued)

5.4.3 *Verification of the Estimation Pore Volume Compressibility* c_p

Sample deformation is measured synchronously by the radial and axial extensometer during the transient tests. Martin and Chandler [35]reported that the stress–strain of rock was at the linear elasticity under low-pressure conditions.

$$\begin{cases} \varepsilon_1 = \dfrac{1}{E}(\sigma_1 - 2\nu\sigma_2)\ (\text{axial}) \\ \varepsilon_2 = \dfrac{1}{E}[\sigma_2 - \nu(\sigma_1 + \sigma_2)]\ (\text{radial}) \end{cases} \tag{5.40}$$

where ε_1 and ε_2 are the axial and radial strains, respectively; σ_1 and σ_2 are the maximum principal stress and the minimum principal stress, respectively. Triaxial compression tests were carried out on the sample under different confining pressures. The axial and radial strains were synchronously measured during the tests. Based on Eq. (5.40) and the axial and radial strains of eight groups of experimental data, E and ν can be fitted by the least-square method. Figure 5.7 shows the curve of axial strains ε_1 versus stresses σ_1. The Red dots in the figure were measured from

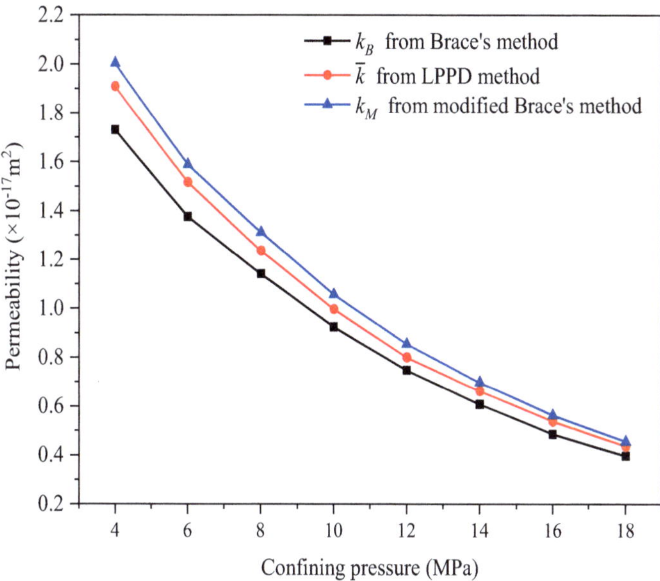

Fig. 5.6 Comparison of permeability calculated by LPPD method, Brace's method, and modified Brace's method

8 groups of transient tests. The solid line is the fitting curve of experimental data through Eq. (5.40). By fitting the experimental data through Eq. (5.40), we can obtain Young's Modulus $E = 2.802 \times 10^9$ Pa, and Poisson's ratio $v = 0.324$.

In the elastic stage, the bulk modulus of sample K can be written as

$$K = \frac{E}{3(1 - 2v)} \tag{5.41}$$

The volume modulus K is reciprocal to the bulk compressibility c_b

$$c_b = \frac{1}{K} \tag{5.42}$$

Combined Eqs. (5.42) with (5.11), the pore volume compressibility c_p can be changed into

$$c_p = \frac{3(1 - 2v)}{\varphi E} \tag{5.43}$$

With the obtained E and v, the pore volume compressibility of the tested sample is calculated as $c_p = 3.332 \times 10^{-8}$ pa^{-1}. This value is close to the estimated pore volume compressibility (as listed in Table 5.2) from the LPPD method. The maximum error between them is -6.93%, demonstrating the applicability of the LPPD method.

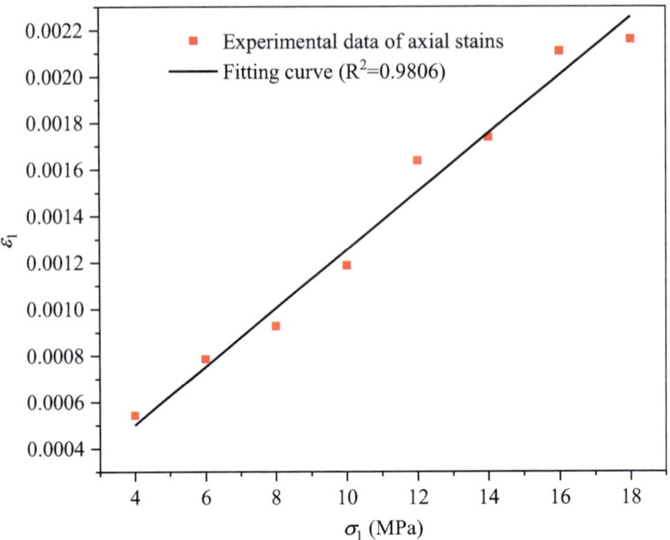

Fig. 5.7 Axial strains of the sample at different stresses

Table 5.2 Estimated permeability and pore volume compressibility under different confining pressure using the LPPD method

P_c (MPa)	k_O (10^{-17} m^2)	c_p (10^{-8} Pa^{-1})
4	1.712	3.441
6	1.362	3.372
8	1.116	3.222
10	0.903	3.127
12	0.725	3.101
14	0.597	3.221
16	0.482	3.446
18	0.394	3.25

After obtaining permeabilities under different confining pressures, the pore volume compressibility can also be determined by fitting the experimental data for permeability-effective stress through Eq. (5.1). Figure 5.8 shows the relationship of permeability and confining pressure under the same pore pressure conditions. By fitting the experimental data in Fig. 5.8, we also obtained the value of pore volume compressibility $c_p = 3.518 \times 10^{-8}$ pa^{-1}, which matches the estimated values in Table 5.2 and the value obtained from Eq. (5.43), further indicating the usefulness of the LPPD method.

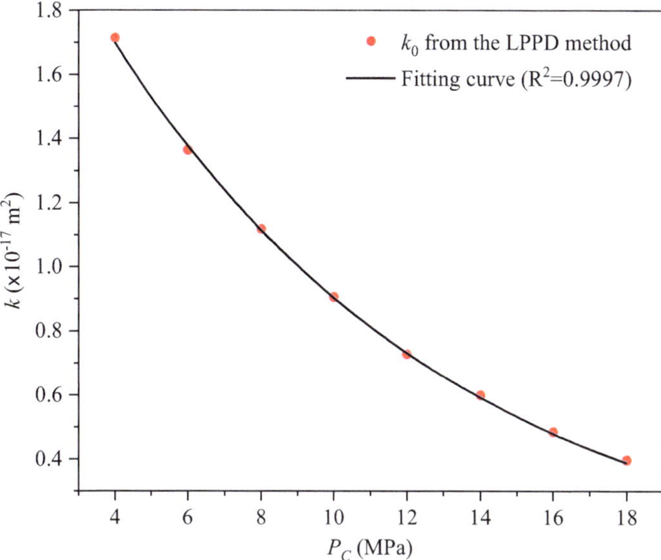

Fig. 5.8 Permeability of the tested sample at different confining pressures

5.4.4 Simplified Method to Estimate k_0 and c_p

Inspired by the findings from the modified method results, for fluids whose compressibility varies little with pore pressure, we proposed a new simpler method for estimating the pore volume compressibility c_p of rocks. Confining pressure possesses a positive effect on the rock permeability, represented by a decrease in value of α in the transient test. Assuming that the slopes of the decay curves under confining pressure P_{C1} and P_{C2} are α_{PC1} and α_{PC1}, respectively. Then the following system of equations can be set up

$$\begin{cases} \alpha_{P_{C1}} &= \dfrac{fAk_{PC1}}{\mu\beta L}\left(\dfrac{1}{V_u}+\dfrac{1}{V_d}\right) \\ \alpha_{P_{C2}} &= \dfrac{fAk_{PC2}}{\mu\beta L}\left(\dfrac{1}{V_u}+\dfrac{1}{V_d}\right) \\ k_{PC2} &= k_{PC1}e^{-3c_p(P_{C2}-P_{C1})} \end{cases} \tag{5.44}$$

The above equations can be transformed into

$$\frac{\alpha_{PC2}}{\alpha_{PC1}} = e^{-3c_p(P_{C2}-P_{C1})} \tag{5.45}$$

Figure 5.9 shows the curves in terms of $\ln[\Delta P(t)/\Delta P(t_0)]$ against time t, the slope α can be calculated from the straight segment of the curve. Table 5.3 lists the pore volume compressibility c_p calculated by the ratios of different slopes under different confining pressures. The estimated values of c_p range from 3.102 to 3.821 pa^{-1}, which are very close to the estimated values in Table 5.2 and the value obtained from Eq. (5.43). The reason for this small deviation is that the method is not an exact analytical solution but a simplified approximation result. However, compared with the value $c_p = 3.332 \times 10^{-8}$ pa^{-1} obtained from Eq. (5.43), the error is 5.25%, indicating that this method is still a simple and feasible method to estimate c_p. It should be noted that the above-estimated pore volume compressibility is obtained under the premise that the pore volume compressibility is constant. It is relatively easy to obtain the slope of the pressure pulse decay curve under different confining pressures. Therefore, the simplified method is a relatively effective and convenient method to estimate the pore volume compressibility of rock samples.

With the obtained value of c_p, we can predicate and calculate the permeability under different stress conditions. For example, after estimating the permeability and pore volume compressibility under the confining pressure of 4 MPa, we plotted the prediction permeability curve of the tested sample in the range of effective stress 0–50 MPa (as shown in Fig. 5.10).

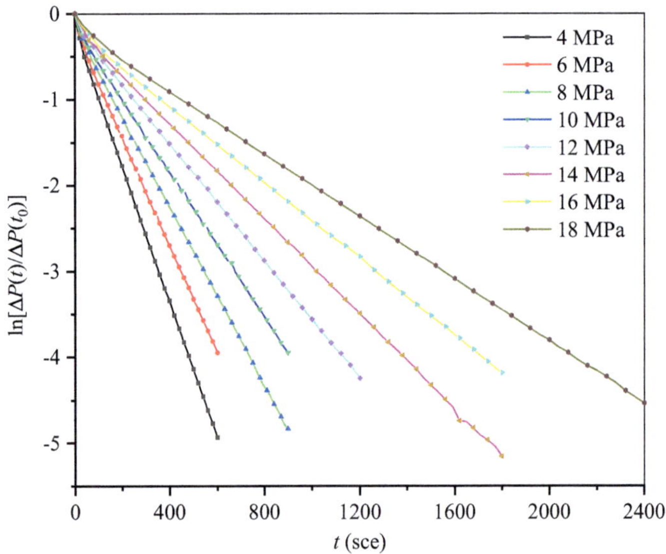

Fig. 5.9 ln[$\Delta P(t)/\Delta P(t_0)$] vs t curve under different confining pressures

Table 5.3 The pore volume compressibility c_p calculated by the ratios of different slopes under different confining pressures (the subscript refers to the confining pressure value)

$\alpha_{PC2}/\alpha_{PC1}$	α_6/α_4	α_8/α_6	α_{10}/α_8	α_{12}/α_{10}	α_{14}/α_{12}	α_{16}/α_{14}	α_{18}/α_{16}
c_p (10^{-8} Pa^{-1})	3.821	3.102	3.537	3.518	3.468	3.713	3.389

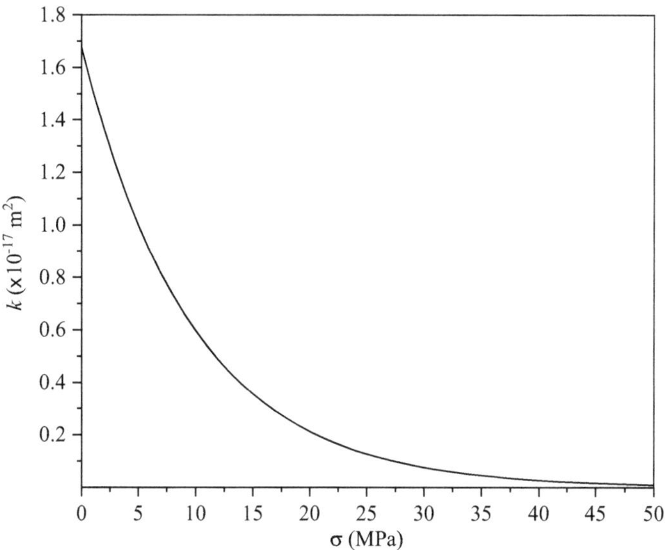

Fig. 5.10 Prediction permeability of the sample under different effective stresses

5.5 Conclusion

An LPPD method is established to estimate the initial permeability k_0 and pore volume compressibility c_p of the rock sample. Due to the complexity of the equations set in the LPPD method, it is very difficult to get the analytical solution directly. Therefore, a numerical solution of the model using the finite difference method is developed. A globally convergent least-square method is proposed to estimate the two unknown parameters (k_0, c_p) in the method by fitting the pressure pulse decay curve. A series of transient tests with large pressure pulses are conducted based on a coal sample under different confining pressure. The fitting results with 8 groups of experimental data under different confining pressures show that the LPPD method can well match the experimental data and facilitate obtaining the k_0 and c_p. Considering the influence of pore volume compressibility, a modified Brace's method is developed for incompressible fluids. The estimated permeability is compared and verified with Brace's method and modified Brace's method, which shows that the LPPD method is accurate and reliable for estimating rock permeability. The estimated pore volume compressibility is verified with experimental results obtained from measured deformation parameters. The comparison results show the consistency and validity of the LPPD method results. Compared with the method of measuring permeability and pore volume compressibility separately, LPPD can measure both of them in one test and effectively shorten the test time. Inspired by the modified Brace's method, for fluids whose compressibility varies little with pore pressure, a more convenient and simple method is proposed to estimate the sample's pore volume compressibility. The new method requires at least two different slopes of pressure decay curves under different confining pressures and is verified by experimental results. It is easy to obtain the slope of the pressure pulse decay curve under different confining pressures. Therefore, the newly presented method is a relatively effective and convenient method to estimate the pore volume compressibility of rock samples.

References

1. Kim TH, Cho J, Lee KS (2017) Evaluation of CO_2 injection in shale gas reservoirs with multi-component transport and geomechanical effects. Appl Energy 190:1195–1206. https://doi.org/10.1016/j.apenergy.2017.01.047
2. Wang C, Pan L, Zhao Y, Zhang Y, Shen W (2019) Analysis of the pressure-pulse propagation in rock: a new approach to simultaneously determine permeability, porosity, and adsorption capacity. Rock Mech Rock Eng 52:4301–4317. https://doi.org/10.1007/s00603-019-01874-w
3. He B, Xie L, Zhao P, Ren L, Zhang Y (2020) Highly efficient and simplified method for measuring the permeability of ultra-low permeability rocks based on the pulse-decay technique. Rock Mech Rock Eng 53:291–303. https://doi.org/10.1007/s00603-019-01911-8
4. Zhao Y, Wang CL, Bi J (2020) Analysis of fractured rock permeability evolution under unloading conditions by the model of elastoplastic contact between rough surfaces. Rock Mech Rock Eng 53:5795–5808. https://doi.org/10.1007/s00603-020-02224-x
5. Cui X, Bustin MM, Bustin RM (2009) Measurements of gas permeability and diffusivity of tight reservoir rocks: different approaches and their applications. Geofluids 9:208–223. https://doi.org/10.1111/j.1468-8123.2009.00244.x
6. Cao C, Li T, Shi J, Zhang L, Fu S, Wang B, Wang H (2016) A New Approach for Measuring the permeability of shale featuring adsorption and ultra-low permeability. J Nat Gas Sci Eng 30:548–556. https://doi.org/10.1016/j.jngse.2016.02.015
7. Kumar H, Elsworth D, Mathews JP, Marone C (2016) Permeability evolution in sorbing media: analogies between organic-rich shale and coal. Geofluids 16:43–55. https://doi.org/10.1111/gfl.12135
8. Sander R, Pan Z, Connell LD (2017) Laboratory measurement of low permeability unconventional gas reservoir rocks: a review of experimental methods. J Nat Gas Sci Eng 37:248–279. https://doi.org/10.1016/j.jngse.2016.11.041
9. Zhao Y, Wang C, Ning L, Zhao H, Bi J (2022) Pore and fracture development in coal under stress conditions based on nuclear magnetic resonance and fractal theory. Fuel 309:122112. https://doi.org/10.1016/j.fuel.2021.122112
10. Brace WF, Walsh JB, Frangos WT (1968) Permeability of granite under high pressure. J Geophys Res 73:2225–2236. https://doi.org/10.1029/JB073i006p02225
11. Hsieh PA, Tracy JV, Neuzil CE, Bredehoeft JD, Silliman SE (1981) A transient laboratory method for determining the hydraulic properties of 'tight' rocks—i. Theory. Int J Rock Mech Min Sci Geomech Abstr 18:245–252. https://doi.org/10.1016/0148-9062(81)90979-7
12. Dicker A, Smits R (1988) A practical approach for determining permeability from laboratory pressure-pulse decay measurements. In: International meeting on petroleum engineering. Society of Petroleum Engineers. https://doi.org/10.2523/17578-MS
13. Jones SC (1997) A technique for faster pulse-decay permeability measurements in tight rocks. SPE Form Eval 12:19–26. https://doi.org/10.2118/28450-PA
14. Wang Y, Nolte S, Tian Z, Amann-Hildenbrand A, Krooss B, Wang M (2021) A modified pulse-decay approach to simultaneously measure permeability and porosity of tight rocks. Energy Sci Eng. https://doi.org/10.1002/ese3.989
15. Hannon MJ (2016) Alternative approaches for transient-flow laboratory-scale permeametry. Transp Porous Med 114:719–746. https://doi.org/10.1007/s11242-016-0741-8

16. Yang Z, Dong M, Zhang S, Gong H, Li Y, Long F (2016) A method for determining transverse permeability of tight reservoir cores by radial pressure pulse decay measurement. J Geophys Res Solid Earth 121:7054–7070. https://doi.org/10.1002/2016JB013173
17. Feng R, Liu J, He Y, Chen S (2020) Fast permeability measurements of tight and sorptive gas reservoirs using a radial-flow transient technique. J Nat Gas Sci Eng 84:103673. https://doi.org/10.1016/j.jngse.2020.103673
18. Liu H, Rutqvist J (2010) A new coal-permeability model: internal swelling stress and fracture-matrix interaction. Transp Porous Med 82:157–171. https://doi.org/10.1007/s11242-009-9442-x
19. Shedid SA, Moustafa EAA (2010) Influence of rock permeability and asphaltene content of crude oil on pore volume compressibility: an experimental approach. Pet Sci Technol 28:27–41. https://doi.org/10.1080/10916460701856633
20. Moosavi SA, Goshtasbi K, Kazemzadeh E, Bakhtiari HA, Esfahani MR, Vali J (2014) Relationship between porosity and permeability with stress using pore volume compressibility characteristic of reservoir rocks. Arab J Geosci 7:231–239. https://doi.org/10.1007/s12517-012-0760-x
21. Oliveira GLPD, Ceia MAR, Missagia RM, Archilha NL, Figueiredo L, Santos VH, Neto LI (2016) Pore volume compressibilities of sandstones and carbonates from helium porosimetry measurements. J Pet Sci Eng 137:185–201. https://doi.org/10.1016/j.petrol.2015.11.022
22. Hou X, Zhu Y, Wang Y, Liu Y (2019) Experimental study of the interplay between pore system and permeability using pore compressibility for high rank coal reservoirs. Fuel 254:115712. https://doi.org/10.1016/j.fuel.2019.115712
23. Hall HN (1953) Compressibility of reservoir rocks. J Pet Technol 5:17–19. https://doi.org/10.2118/953309-G
24. Zhu S, Du Z, Li C, You Z, Peng X, Deng P (2018) An analytical model for pore volume compressibility of reservoir rock. Fuel 232:543–549. https://doi.org/10.1016/j.fuel.2018.05.165
25. Sui W, Quan Z, Hou Y, Cheng H (2020) Estimating pore volume compressibility by spheroidal pore modeling of digital rocks. Pet Explor Dev 47:603–612. https://doi.org/10.1016/S1876-3804(20)60077-5
26. Feng R, Liu J, Harpalani S (2017) Optimized pressure pulse-decay method for laboratory estimation of gas permeability of sorptive reservoirs: part 1—background and numerical analysis. Fuel 191:555–564. https://doi.org/10.1016/j.fuel.2016.11.079
27. Zhao Y, Hu Y, Wei J, Yang D (2003) The experimental approach to effective stress law of coal mass by effect of methane. Transp Porous Med 53:235–244. https://doi.org/10.1023/A:1025080525725
28. Connell LD (2016) A new interpretation of the response of coal permeability to changes in pore pressure, stress and matrix shrinkage. Int J Coal Geol 162:169–182. https://doi.org/10.1016/j.coal.2016.06.012
29. Zhao Y, Wang Y, Wang W, Tang L, Liu Q, Cheng G (2019) Modeling of rheological fracture behavior of rock cracks subjected to hydraulic pressure and far field stresses. Theor Appl Fract Mech 101:59–66. https://doi.org/10.1016/j.tafmec.2019.01.026
30. Mckee CR, Bumb AC, Koenig RA (1988) Stress-dependent permeability and porosity of coal and other geologic formations. SPE Form Eval 3:81–91. https://doi.org/10.2118/12858-PA
31. Somerton WH, Söylemezoḡlu IM, Dudley RC (1975) Effect of stress on permeability of coal. Int J Rock Mech Min Sci Geomech Abstr 12:129–145. https://doi.org/10.1016/0148-9062(75)91244-9
32. Yang Z, Wang W, Dong M, Wang J, Li Y, Gong H, Sang Q (2016) A model of dynamic adsorption-diffusion for modeling gas transport and storage in shale. Fuel 173:115–128. https://doi.org/10.1016/j.fuel.2016.01.037
33. Ezquerro JA, Hernández-Verón MA (2020) A new concept of convergence for iterative methods: restricted global convergence. J Comput Appl Math 113051. https://doi.org/10.1016/j.cam.2020.113051

34. Yano M, Huang T, Zahr MJ (2021) A globally convergent method to accelerate topology optimization using on-the-fly model reduction. Comput Method Appl Mech Eng 375:113635. https://doi.org/10.1016/j.cma.2020.113635
35. Martin CD, Chandler NA (1994) The progressive fracture of Lac du Bonnet granite. Int J Rock Mech Min Sci Geomech Abstr 31:643–659. https://doi.org/10.1016/0148-9062(94)90005-1

Chapter 6
A Canister Degassing Method for Simultaneous Determination of Axial and Transverse Permeability in Tight Reservoir Cores

6.1 Introduction

Gas extraction from unconventional reservoirs such as shale and tight gas reservoirs has substantially increased fossil-energy production in countries such as USA, Canada, and China. Permeability is one of the most critical reservoir parameters for shale and tight gas reservoirs characterization and well-performance evaluation because it affects the production rate, the pace of recovery and the technically recoverable hydrocarbons [1]. However, an accurate measurement of shale and tight reservoirs is challenging due to the low/ultra-low permeability and heterogeneous nature of shale formations.

A lot of experimental techniques have been developed to measure shale and tight gas reservoirs permeability. These techniques can be divided into two categories: steady-state and unsteady-state. The steady-flow technology is usually not adequate for shale and tight gas reservoirs because it requires a long time to reach the steady-state flow regime. Therefore, the unsteady-state technique becomes the preferred option. The most widely used unsteady-state technique are pressure pulse decay method [2–5], oscillating pulse method [6–8], canister degassing test method [1, 9, 10], and the Gas Research Institute (GRI) method [11, 12]. These methods have several advantages: (1) short experimental duration, and (2) the pressure signal or the cumulative flow as opposed to instantaneous flow rate is measured. With these proposed methods, the effects of stress [13], fracture properties [14], water content [15], and gas adsorption [16] on shale permeability are investigated. However, since these permeability measurement methods are developed by assuming one-dimension flow, only axial permeability of a core sample or spherical permeability of crushed samples can be obtained. Therefore, most of these studies are limited to homogenous rock. It is, in fact, important to develop a method to measure the anisotropic permeability because shale and tight gas reservoirs are indeed anisotropic rocks. The permeability anisotropy, defined as the ratio between permeability parallel to bedding and perpendicular to bedding, is generally greater than one, but varies greatly from 1.2

C. Wang and Y. Zhao, *Permeability Measurement of Tight Rock*,
https://doi.org/10.1007/978-981-92-0597-4_6

to 1864.4 for Longmaxi and Wufeng shales from the Sichuan Basin in South China [17]. The large variation in permeability anisotropy can be attributed to the composition, grain size, porosity, effective stress, and micro-fractures [18–20]. Mokhtari and Tutuncu [21] tested directional (vertical, 45-degree, and horizontal) Mancos shale in Colorado using unsteady-state techniques and showed that the presence of lamination and fractures caused significant directional dependency of permeability.

Due to the lack of appropriate method to measure permeability anisotropy, the most used method is to drill two test cores parallel to bedding and perpendicular to bedding respectively to measure their permeabilities using unsteady-state techniques. However, the artificial micro-fractures of different core samples cause this method to be unreliable. A huge error may exist because the permeability measurement in different directions cannot be based on the same core sample [22].

In this section, an experimental method for simultaneously determining axial and transverse permeability of tight reservoir cores is developed. The experiments, known as the canister degassing test, are originally performed for calculation of lost gas by Hosseini et al. [23] and are further improved by Alfi et al. [10] for permeability measurement on homogenous rock. We modified their work to simultaneously determine axial and transverse permeability of core samples. The method is verified through numerical simulations and is successfully applied to Longmaxi shale core to estimate the axial and transverse permeability. At last, the influence factors and limitations of the proposed method are discussed.

6.2 Method

The core samples are saturated with gas at initial gas pressure first and then degasified inside a canister at a lower constant pressure. Gas production data are recorded by the flowmeter for calculation of axial and transverse permeability. All sides of the core sample (top, bottom, and radial surrounding surfaces) are open to flow at constant pressure.

Figure 6.1 shows the schematic model of canister degassing tests. The apparatus is composed of a canister, a high-resolution flowmeter (made from OMEGA company, USA, which can measure gas flow rate from 0 to 10 mL per min, with precision of 0.01 mL), pressure transducer (made from OMEGA company, which can measure gas pressure of 0 to 100 MPa, with precision of 0.001 MPa), back pressure regulator, gas supplement system (consists of gas cylinder, air compressor, and gas chamber, which can provide gas pressure of 0 to 20 MPa), and vacuum system.

The procedure of canister degassing tests is summarized as: (1) Preparation, place the sample into the canister, and vacuumize the system for 24 h. (2) Gas saturation, open valve 2 to bring the canister a required pressure. The core is saturated with Helium to achieve the desired equilibrium pressure. (3) Gas drawdown, a lower constant pressure (or atmospheric pressure) is applied at the outer boundary to let the core gas release through a back-pressure regulator. More details of the canister degassing tests can be found in Refs. [1, 10].

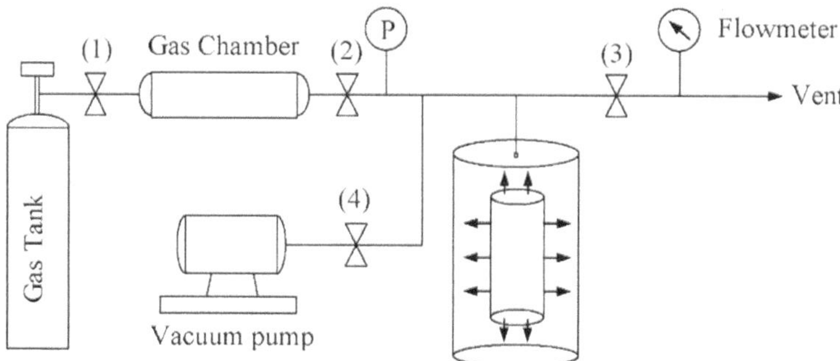

Fig. 6.1 Schematic model of canister degassing apparatus. The sample is saturated with gas at initial gas pressure first and then degasified inside a canister at a lower constant pressure. The cumulative volume of degassed gas is chronologically recorded by the flowmeter for the permeability measurement

Two core samples, drilled from an outcrop of the lower Silurian Longmaxi formation in the Sichuan Basin, China, were used for permeability tests. Figure 6.2 shows the prepared samples, which are 50 mm in length and 50 mm in diameter. To investigate the permeability anisotropy, one sample (namely vertical sample) is drilled parallel to the bedding, and another (namely horizontal sample) is perpendicular to bedding. The mineral compositions of the samples are quartz, potassium feldspar, albite, illite, chlorite, Calcite, and Pyrite, The total organic carbon (TOC) of the samples is 2.88%, and vitrinite reflectance (Ro) is 3.72%.

Fig. 6.2 Prepared shale samples (The left is vertical sample, the right is horizontal sample)

6.2.1 Analytical Solution

The model is developed based on the single-phase gas flow in a radially and vertically finite core sample, considering the effect of transient flow with pseudopressures [23]. For the case of a heterogeneous cylinder core, the mass-continuity equation can be written as:

$$\frac{\partial(\phi\rho_g)}{\partial t}=\frac{1}{r}\frac{\partial(r\rho_g v_r)}{\partial r}+\frac{\partial(\rho_g v_x)}{\partial x} \tag{6.1}$$

where ϕ is porosity, ρ_g is gas density, t is time, v_r and v_x refer to the flow velocities in r direction and x direction, respectively. Note that gas adsorption is not considered in this mass-continuity equation.

To account for changes in the compressibility and viscosity of gas caused by pressure decline, the pseudopressure is used as [24]:

$$m(p)=2\int_{p_i}^{p}\frac{p}{\mu z}dp \tag{6.2}$$

Based on the equation of state, the gas density is:

$$\rho_g=\frac{pM}{zR_0T} \tag{6.3}$$

where M is molecular weight of gas, R_0 universal gas constant (8.314 J K^{-1} mol^{-1}), T is experimental temperature.

Assuming that Darcy's law prevails, gives:

$$v_r=\frac{k_r}{\mu}\frac{\partial p}{\partial r} \tag{6.4}$$

$$v_x=\frac{k_x}{\mu}\frac{\partial p}{\partial x} \tag{6.5}$$

where k_r and k_x are the permeabilities in r direction and x direction, respectively.

Replacing these parameters give

$$\frac{\phi\mu c}{k_r}\frac{\partial m(p)}{\partial t}=\frac{1}{r}\frac{\partial\left[r\frac{\partial m(p)}{\partial r}\right]}{\partial r}+\frac{k_x}{k_r}\frac{\partial\left[\frac{\partial m(p)}{\partial x}\right]}{\partial x} \tag{6.6}$$

The boundary and initial conditions are as follows:

$$\begin{aligned} m_p(r=R,x,t>0) &= m_{p0} \\ m_p(r,x=L,t>0) &= m_{p0} \end{aligned}$$

$$
\begin{aligned}
m_p(r, x = 0, t > 0) &= m_{p0} \\
\lim_{r \to 0} r \frac{\partial m(p)}{\partial r} &= 0 \\
m_p(r, x, 0) &= m_{pi}
\end{aligned}
\tag{6.7}
$$

where L and R denote length and radius of the core sample. m_{p0} is the pseudopressure at the equivalent pressure, and m_{pi} is the initial gas pseudo-pressure in the core sample.

For convenience, a set of dimensionless parameters are defined as

$$
\begin{aligned}
r_D &= \frac{r}{R} \\
x_D &= \frac{x}{L} \\
t_D &= \frac{k_r t}{\phi \mu c R^2} \\
m_{pD} &= \frac{m_p - m_{p0}}{m_{pi} - m_{p0}}
\end{aligned}
\tag{6.8}
$$

Replacing for the dimensionless parameters yield:

$$
\frac{\partial m_{p_D}}{\partial t_D} = \frac{\partial^2 m_{p_D}}{\partial r_D^2} + \frac{1}{r_D}\frac{\partial m_{p_D}}{\partial r_D} + \nu^2 \frac{\partial^2 m_{p_D}}{\partial x_D^2}
\tag{6.9}
$$

where ν^2 is defined as

$$
\nu^2 = \chi \frac{R^2}{L^2}, \ \chi = \frac{k_x}{k_r}
\tag{6.10}
$$

The solution of Eq. (6.9) with corresponding boundary and initial conditions for dimensionless pressure is given by [23]:

$$
m_{pD}(r_D, x_D, t_D) = 2\sum_{m=1}^{\inf}\left\{\left[2\sum_{m=1}^{\inf}\left\{\left[e^{-t_D\left(\omega_m^2 \nu^2 + \lambda_n^2\right)}\right] \times \frac{[1-\cos(\omega_m)]F(\lambda_n)}{\omega_m}\right\}\frac{J_0(\lambda_n r_D)}{|J_1(\lambda_n)|^2}\right]\sin(\omega_m x_D)\right\}
\tag{6.11}
$$

where λ_n is defined as

$$
\lambda_n = J_0(0, n), \ n = 1, 2, 3\ldots
\tag{6.12}
$$

the roots of λ_n are of the zero-order, first-kind Bessel function:

$$
\lambda_1 = 2.40483, \lambda_2 = 5.52008, \lambda_3 = 8.65373, \lambda_4 = 11.7915, \lambda_5 = 14.9309
$$

and $F(\lambda_n)$ is

$$F(\lambda_n) = \frac{J_1(\lambda_n)}{\lambda_n} \tag{6.13}$$

With the pressure profiles, gas produced from the sides of the core and the top–bottom portion can be calculated as:

$$\begin{aligned} q_s(t_D) &= \lim_{r \to R} -\frac{k_r A}{\mu}\frac{\partial p}{\partial r} \\ q_{tb}(t_D) &= 2\lim_{x \to 0} -\frac{k_x A}{\mu}\frac{\partial p}{\partial x} \end{aligned} \tag{6.14}$$

where q_s and q_{tb} are the gas flow from the sides of the core and the top–bottom portion, respectively. Adding these terms gives the total gas production from the core sample as:

$$\begin{aligned} Q(t) =& Q_s(t) + Q_{tb}(t) \\ =& \frac{4zV_c\eta\left(m_{pi} - m_{po}\right)}{p}\sum_{m=1}^{\inf}\sum_{n=1}^{\inf}\left[\frac{1}{\left(\omega_m^2 \nu^2 + \lambda_n^2\right)}\left(1 - e^{-\frac{t\left(\omega_m^2\nu^2+\lambda_n^2\right)}{\eta R^2}}\right)\right] \times \\ & \left[k_r\frac{[1-\cos(\omega_m)]^2}{\omega_m^2} + 2k_x\frac{\nu^2[1-\cos(\omega_m)]}{\chi\lambda_n^2}\right] \end{aligned} \tag{6.15}$$

where V_c is the volume of the core, η is the inverse of diffusivity coefficient, ω_m is the Fourier transform eigenvalues, λ_n Hankel-transform eigenvalue, χ is the ratio of axial permeability to transverse permeability.

The gas production is analyzed by history matching the analytical solution to the measured test data. In the history-matching process, there are only two unknowns: transverse permeability and permeability ratio. All other parameters, including core dimensions and gas properties are assumed to be known. Generally, parameter estimation problems can be transformed into least-squares problems. Many optimization methods are specially designed to solve least-squares problems developed from Newton's method. One very popular local optimization method is the Levenberg–Marquardt (LM) method. However, the LM method alone is not sufficient to solve our problem because Eq. (6.15) is non-convex problem, which means that it has many local minima and that the global minimum is essentially impossible to find solely by a local optimization method. Therefore, a global optimization method should be used to solve this problem. The global optimization method used in this section is a multilevel single-linkage (MLSL) method. It combines a stochastic method with a local optimization method, which has proved to be a very effective heuristic means of solving very complex global optimization problems. Therefore, axial and transverse permeability of the core sample can be simultaneously determined with CDT method. We should note that the obtained permeabilities are apparent permeabilities if slippage effects exist. We mainly focus on the method for determining axial and transverse permeability through CDT method. The absolute permeabilities can be obtained by a sets of permeability tests under different gas pressure. Interested readers may refer to [25] for the details.

6.2.2 *Verification of the Proposed Method*

To verify the accuracy of the proposed method, we performed numerical simulations of 3 cases that cover variations in permeabilities. The axial and transverse permeability were calculated using the method described in Sect. 2.1, from the simulated gas production decay data, and then compare them to the values actually used in the numerical simulations.

In the numerical simulations, the core sample was discretized into 2500 grid cells (shown in Fig. 6.3) and the canister was represented as a special grid cell attached to the sides of the core. The sample dimensions are representative of actual core tested in the laboratory flow tests, that is, 50 mm in diameter and 50 mm in length. These simulations were completed using TOUGH + REALGASBRINE (TOUGH+), a widely used numerical simulator for non-isothermal multiphase flow (an aqueous phase and a real gas mixture) in a gas-bearing medium, with a particular focus in ultra-tight systems [26]. In the numerical tests, the temperature was set to 25 °C, initial pressure and gas drawdown pressure were set to 3 MPa and 2 MPa, respectively.

Figure 6.4 shows the gas flow data generated from three numerical tests. With these data, we calculated the axial and transverse permeability by history match. Table 6.1 shows the comparison between the calculated permeability using the proposed method from the data generated by numerical simulations and the true values used in numerical simulation. As shown in Table 6.1, the determined axial and transverse permeability values are very close to the true values. The relative error between the mean estimated value and the true value ranges from 1.7% to 4.5% for transverse permeability, and 1.9% to 5.3% for axial permeability, supporting the usefulness of our method. Note that there are small differences between the results from numerical simulations and analytical solution. This can be explained as follows. In the analytical solution, cumulative gas flow is calculated by the integral of gas flow from an infinitesimal area on the cylinder's periphery, but the simulator is not capable of modeling his process exactly. If the number of grids is infinity, numerical simulation should reproduce the analytical-solution results perfectly. This phenomenon is also observed in Hosseini's numerical tests [23], which indicates that numerical-simulation results get closer to the analytical solution with the increase of grid number.

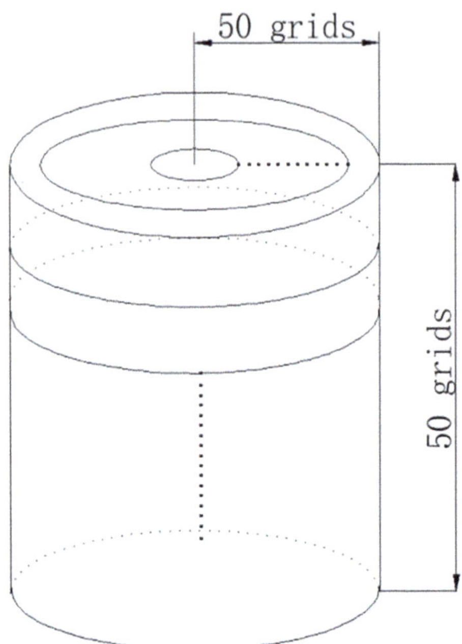

Fig. 6.3 Schematic diagram of numerical simulation model

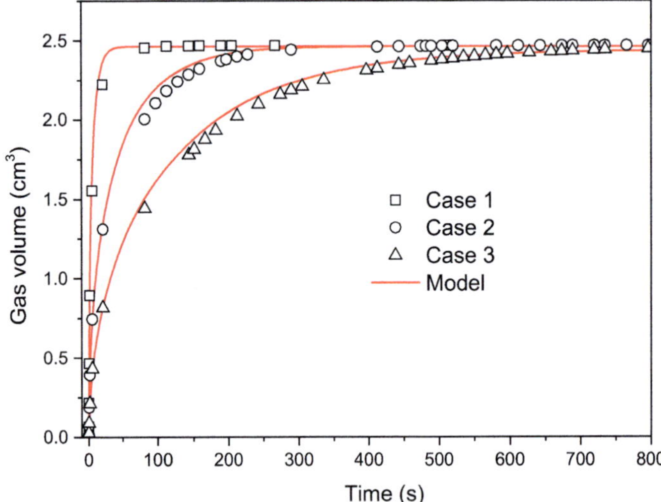

Fig. 6.4 Numerical experimental results

Table 6.1 Numerical cases and calculated results

Numerical tests	True permeability (m^2)		Calculated permeability (m^2)			
	k_r	k_x	k_r	Error (%)	k_x	Error (%)
Case 1	1×10^{-17}	1×10^{-18}	9.83×10^{-18}	1.7	9.69×10^{-19}	3.1
Case 2	1×10^{-18}	1×10^{-18}	9.55×10^{-19}	4.5	9.47×10^{-19}	5.3
Case 3	1×10^{-19}	1×10^{-18}	9.72×10^{-20}	3.8	9.81×10^{-18}	1.9

6.3 Experimental Results

6.3.1 Experimental Results of CDT

Figure 6.5 shows the gas production data versus time for the two samples. From the figure, the gas production increases sharply at the early stage due to the pressure gradient difference. The speed at which gas production increases slows down with flow time. Eventually, gas production stabilizes after the pressure attains the equilibrium pressure. The larger the initial pressure, the more the gas production. We then obtained the axial and transverse permeability by history matching, as shown in Fig. 6.6. Note that the accuracy of the fitting curves decreased with the increase of pressure. This effect arises from the adiabatic changes in temperature caused by changing pressure. Because all the CDT tests are conducted under room temperature and the gas drawdown pressure is set to 1 MPa, the increase of initial gas pressure will cause greater changes in temperature, which further influence the gas properties, leading to a larger fitting deviation. Therefore, to eliminate the effect of temperature, the difference between the initial gas pressure and the gas drawdown pressure should be small.

Figure 6.6 shows that all the permeability values parallel to bedding are bigger than the permeability values perpendicular to bedding. For example, the axial permeability of vertical sample is about 6 times of the transverse permeability, while this ratio decrease to about 0.2 for the horizontal sample. The permeability anisotropy calculated using our method lays very well within the permeability anisotropy values experimentally measured for the same formation and zone [17].

6.3.2 Comparison Between CDT Results and PPD Results

Pressure pulse decay (PPD) method is widely used for axial permeability measurement of tight reservoir rocks. To compare with CDT results, we performed three groups using the PPD method with the core samples. Figure 6.7 shows the schematic sketch of the PPD technique [27–30]. The sample is first in equilibrium and then open the valve between the upstream chamber and the sample as well as the valve

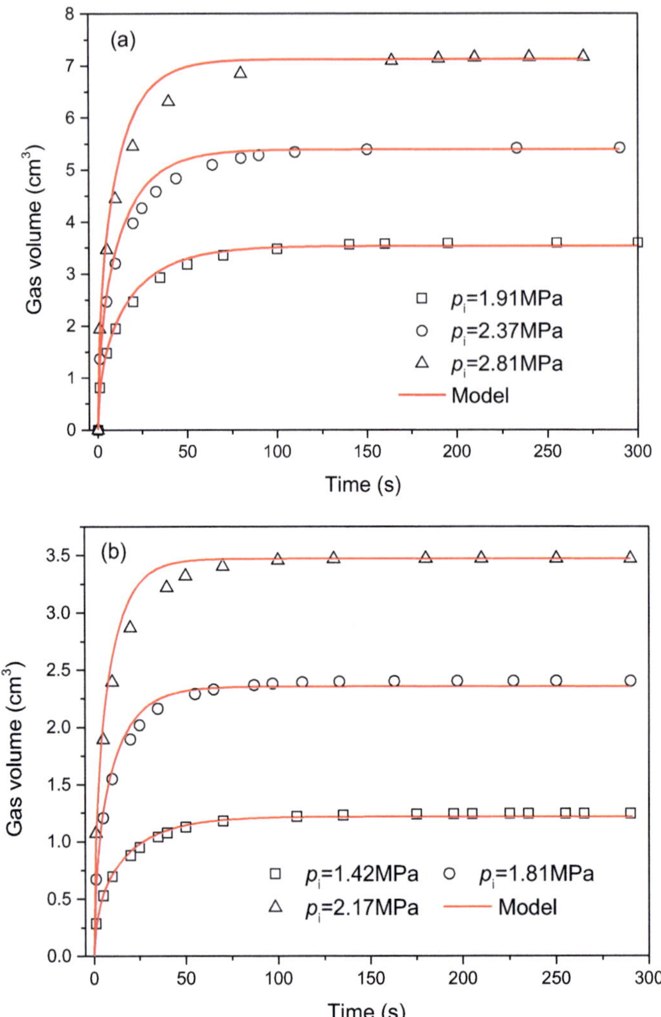

Fig. 6.5 Experimental results of CDT **a** for the vertical sample, **b** for the horizontal sample

between the sample and the downstream chamber to let fluid flowing from the upstream chamber to the downstream chamber through the sample driven by the pressure difference. The axial permeability can then be estimated from the pressure pulse decay curve (the upstream–downstream pressure difference with time) using mathematical solutions of the related flow problem.

Brace's solution is used to estimate axial permeability in PPD tests [2]. Table 6.2 shows the results of PPD tests on the tested samples. Note that transverse permeability cannot be obtained from the PPD method. Therefore, we only compare the axial permeability of the two methods. Compared to the permeabilities measured from

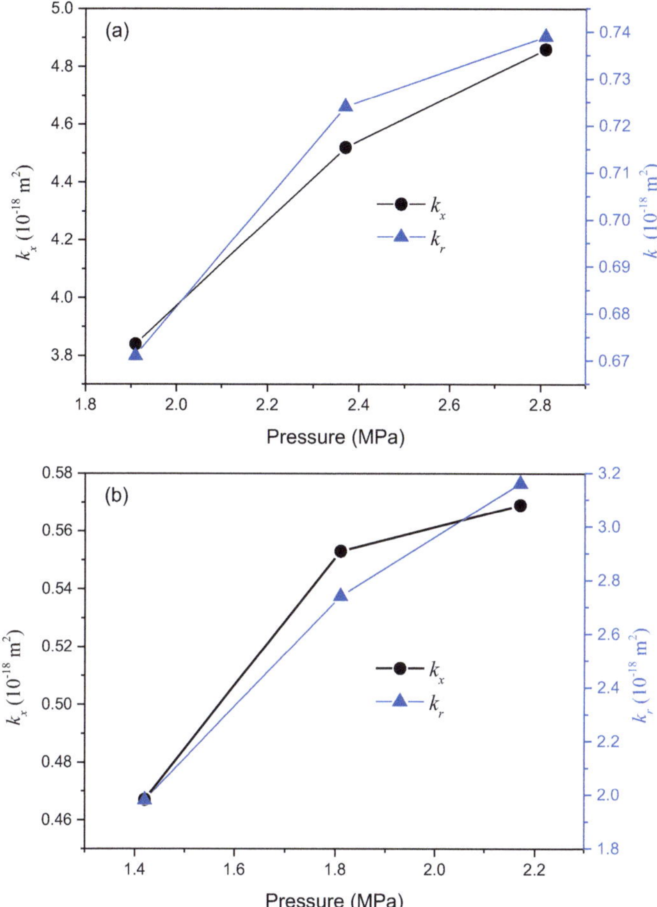

Fig. 6.6 Experimental permeabilities obtained from CDT tests **a** for the vertical sample, **b** for the horizontal sample

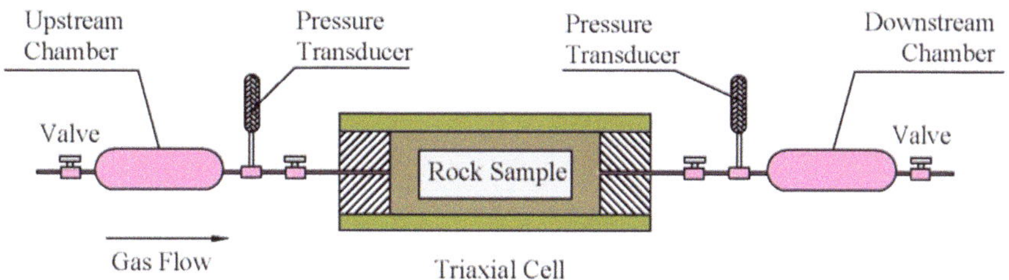

Fig. 6.7 The schematic sketch of the pressure pulse decay technique [27–30]. The sample is first in equilibrium and then open the valve between the upstream chamber and the sample to let fluid flowing from the upstream chamber to the downstream chamber through the sample driven by the pressure difference

Table 6.2 Experimental results of PPD tests on the vertical sample

Confining pressure	Initial equilibrium pressure	Pressure pulse (MPa)	Vertical sample	Horizontal sample
			k_x	k_x
$p_c = 8$ MPa	$p_0 = 2$ MPa	0.1	9.23×10^{-19} m^2	1.47×10^{-19} m^2
$p_c = 8$ MPa	$p_0 = 2.5$ MPa	0.1	1.46×10^{-18} m^2	1.96×10^{-19} m^2
$p_c = 8$ MPa	$p_0 = 3$ MPa	0.1	1.72×10^{-18} m^2	2.24×10^{-19} m^2

CDT tests, the evolution of axial permeability versus gas pressure for the two methods has the same trend (shown in Fig. 6.8). Specifically, permeability increases with the initial gas pressure. Axial permeability values obtained from the PPD method are less than permeability values of CDT method for the full range of gas pressure. The differences existing in the results of the two methods result from the following two reasons. First, PPD tests must be conducted under confining pressure to avoid lateral flow between the rock and sleeve, which may close the micro fractures in the rock and hence decrease the rock permeability. Second, though the initial gas pressures are almost the same for the two methods, the final equilibrium pressure are varied with each other, which causes different slippage effects on the measured results. Meanwhile, we estimated the permeability anisotropy using the axial permeability of the vertical and horizontal samples obtained from PPD tests. The permeability anisotropy values are 6.28, 7.45, and 7.68 respectively, slightly larger than the values obtained from CDT tests.

6.4 Parametric Study and Discussion

In this section, the effects of permeability anisotropy, porosity, initial gas pressure, rock dimensions, and adsorption on the profiles of gas production are investigated. The source of error and limitation of the proposed method are also explored. In the parametric study, we set up a base case and the parameters are: $k_x = k_r = 1 \times 10^{-18}$ m^2, $\phi = 5\%$, $p_i = 1.5$ MPa, $L = 50$ m, $R = 25$ mm.

6.4.1 Effects of Permeability

The curves of cumulative gas volume versus time with variable permeabilities are shown in Fig. 6.9 for cases $k_x = k_r$, and Fig. 6.10 for cases k_x and k_r are varied with each other. The figures clearly illustrate that gas volume produced in the canister is very sensitive to variation in transverse and axial permeability. Both the transverse

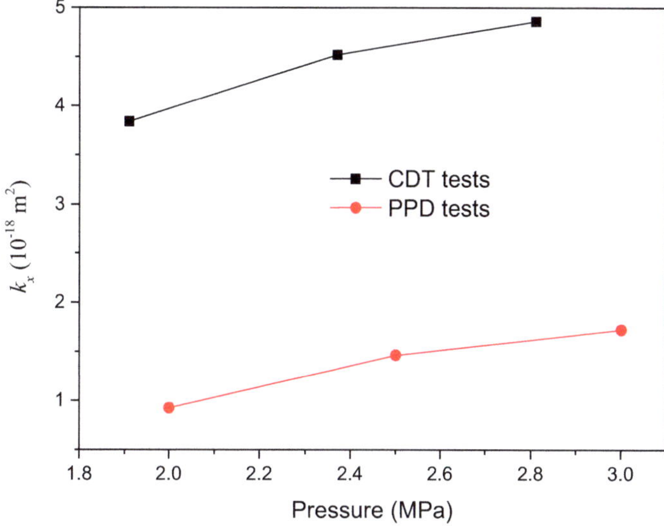

Fig. 6.8 Comparison of axial permeability obtained from CDT tests and PPD tests for vertical sample

and axial permeability affect the gas-flow transient time significantly, specifically, the lower the permeabilities are, the longer the gas-flow transient time is. However, these permeabilities have no effect on the cumulative gas volume.

For cases k_x and k_r have different values, as shown in Fig. 6.10, k_r plays a more important role than k_x in the evolution of cumulative gas volume versus time curve.

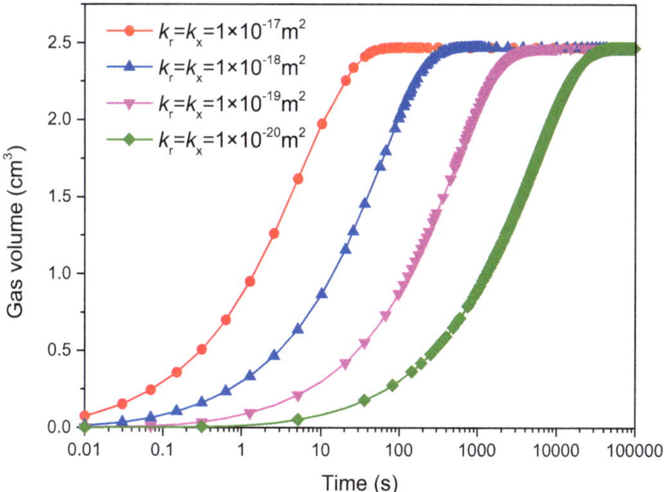

Fig. 6.9 Effects of permeability on CDT testing results when kx is equal to kr

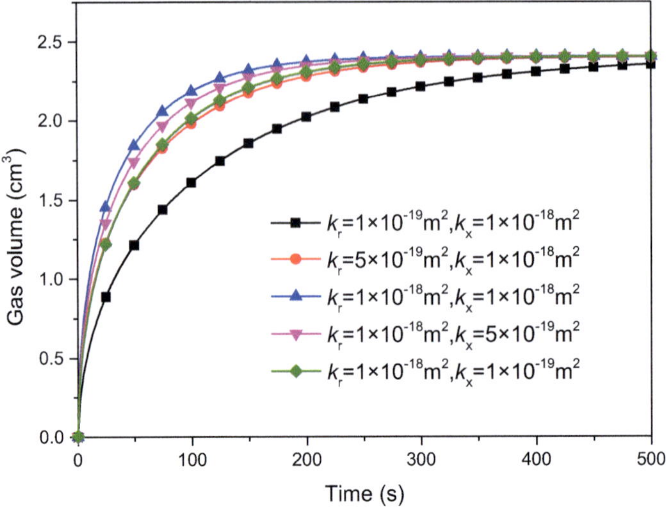

Fig. 6.10 Effects of permeability on CDT testing results when kx differs from kr

For example, the gas-flow transient time of the cases for k_r ranging from 1×10^{-18} to $1 \times 10^{-19}\text{m}^2$ increases much more than the cases for k_x ranging from 1×10^{-18} to $1 \times 10^{-19}\text{m}^2$. Due to that the cumulative gas volume is affected by both permeability and surface area of the core sample, the effects of k_r and k_x are also influenced by the aspect ratio of the core sample (ratio of the sample length to the diameter). When the aspect ratio is much less than 1 (i.e. the sample length is less than the sample diameter), k_x has more contributions to the cumulative gas volume than k_r as a result of larger top–bottom surface area. With the same core sample volume, when the aspect ratio is much larger than 1 (i.e. the sample length is larger than the sample diameter), k_r has more contributions to the cumulative gas volume than k_x as a result of larger lateral surface area.

6.4.2 *Effects of Porosity*

The effects of porosity on the curve of cumulative gas produced versus time is shown in Fig. 6.11. It is apparent that a large pore volume can store more gas and hence affect the gas-flow transient time. As shown in the figure, with the increase of porosity, gas production increases significantly, and the time required to attain final pressure equilibrium is also significantly extended.

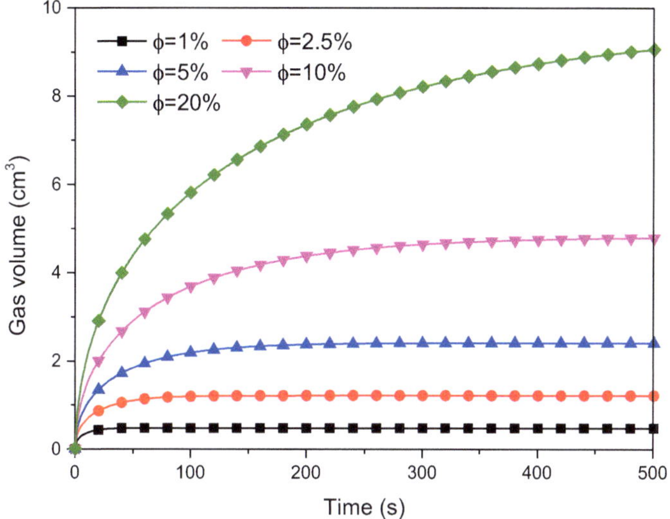

Fig. 6.11 Effects of porosity on CDT tests

6.4.3 Effects of Initial Gas Pressure

Figure 6.12 shows the effects of the initial gas pressure on the curves of cumulative gas volume versus time. As one would intuitively expect, a large initial gas pressure leads to more gas production but has no significant effect on gas-flow transient time. While the initial gas pressure increases from 1.5 to 2.5 MPa, 3.5 MPa, 4.5 MPa, and finally to 5.5 MPa, the final cumulative gas produced increases from 2.4 to 7.26 cm^3, 12.17 cm^3, 17.14 cm^3 and finally to 22.14cm^3, showing a linear relationship.

6.4.4 Effects of Rock Dimensions

To investigate the effects of rock dimensions, sample length and radius are investigated, and the results are shown in Fig. 6.13. The figures indicate that larger rock size will increase both the cumulative gas produced and gas-flow transient time. In particular, the final cumulative gas produced shows a linear relation with sample length and quadratic function relation with sample radius, which indicates a linear relationship between the final cumulative gas produced and sample volume. Therefore, it is advisable, in the interest of more gas production, to use a larger sample when testing a sample with small porosity.

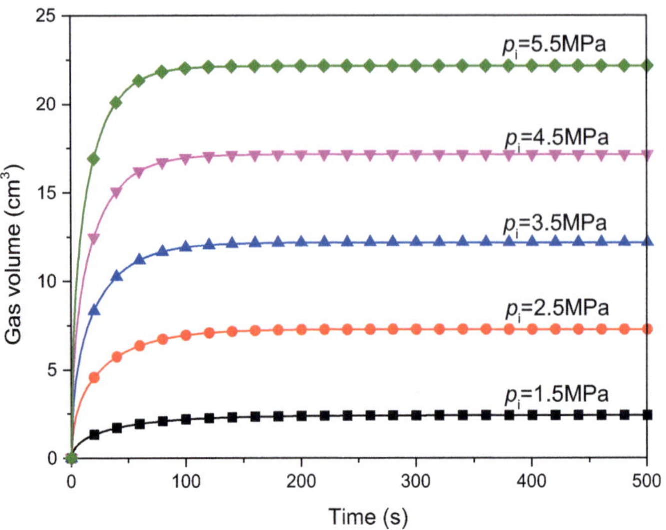

Fig. 6.12 Effect of initial gas pressure on CDT tests

6.4.5 *Effect of Adsorption*

Gas adsorption provides additional storage in the core sample. Therefore, desorbed gas from canister degassing tests should be taken into account if adsorbing gas (e.g. CH_4 and CO_2) are used as testing gases for shale permeability measurement. Equation (6.1) can be revised as follow when accounting for desorbed gas [23]:

$$\frac{\partial(\phi\rho_g)}{\partial t} = \frac{1}{r}\frac{\partial(r\rho_g v_r)}{\partial r} + \frac{\partial(\rho_g v_x)}{\partial x} + \rho_g D \tag{6.16}$$

where D is gas desorption source. To make the final partial differential equation analytically solvable, a linear adsorption relationship is adopted here. This assumption is reasonable at low pressures [10].

$$\psi = \gamma p \tag{6.17}$$

where ψ is the amount of gas desorption, γ is linear adsorption parameter, which can be estimated from actual Langmuir adsorption parameters.

$$\gamma = \frac{d\psi}{dp} = \frac{V_m p_L}{(p + p_L)^2} \tag{6.18}$$

where V_m and p_L are Langmuir volume and pressure constants. Based on the linear adsorption assumption, D can be calculated as:

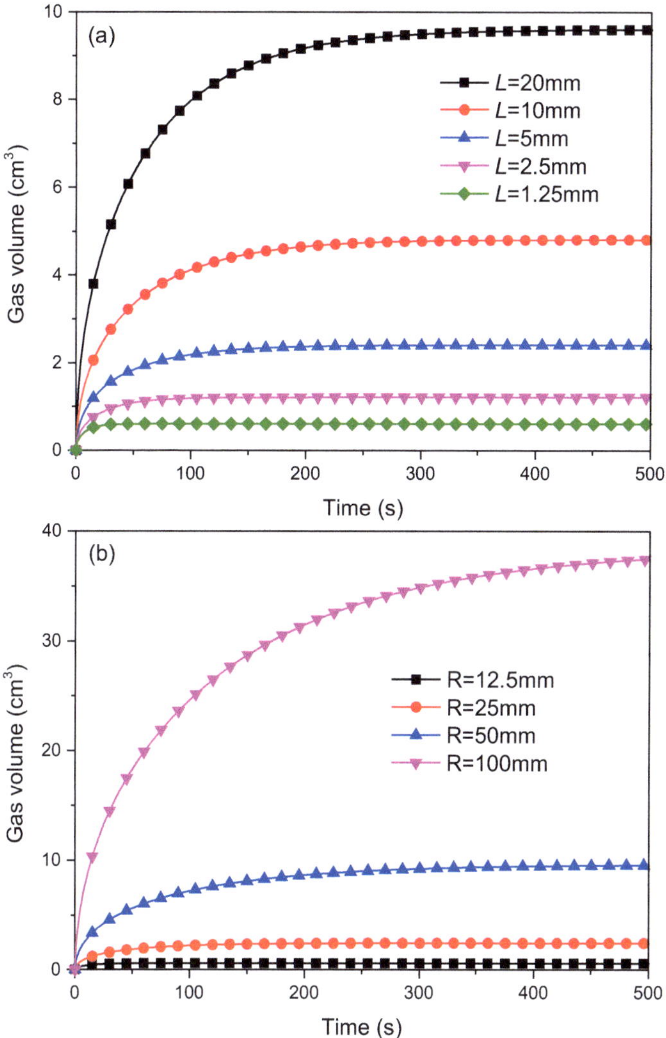

Fig. 6.13 Effect of rock dimensions on CDT tests: **a** for sample length and **b** for sample radius

$$D = -\gamma \rho_s \frac{dp}{dt} \tag{6.19}$$

where ρs is rock density.

Substituting Eqs. (6.19) into (6.16), and replacing the dimensionless parameters, we obtain:

$$(\varpi + 1)\frac{\partial m_{p_D}}{\partial t_D} = \frac{\partial^2 m_{p_D}}{\partial r_D^2} + \frac{1}{r_D}\frac{\partial m_{p_D}}{\partial r_D} + \nu^2 \frac{\partial^2 m_{p_D}}{\partial x_D^2} \tag{6.20}$$

where $\varpi = \frac{\gamma \rho_s}{\phi c}$

The boundary and initial conditions are the same to Eq. (6.7). With the same analytical solving method to Sect. 6.2.2, we obtain the total gas production coupled with gas desorption as:

$$Q(t) = \frac{4zV_c\eta\left(m_{pi} - m_{po}\right)}{p} \sum_{m=1}^{\inf} \sum_{n=1}^{\inf} \left[\frac{(\varpi + 1)}{\left(\omega_m^2 \nu^2 + \lambda_n^2\right)} \left(1 - e^{-\frac{t\left(\omega_m^2 \nu^2 + \lambda_n^2\right)}{\eta R^2 (\varpi + 1)}}\right) \right] \times \left[k_r \frac{[1 - \cos(\omega_m)]^2}{\omega_m^2} + 2k_x \frac{\nu^2 [1 - \cos(\omega_m)]}{\chi \lambda_n^2} \right] \tag{6.22}$$

Figure 6.14 shows the gas production curves with and without consideration of gas desorption ($p_L = 2$ MPa, $V_m = 2.8$ cm^3/g). As expected, the amount of gas production with gas desorption is much larger than the case without gas desorption. The difference is more significant for rocks with larger adsorption capacity. Meanwhile, due to the effect of gas desorption, the gas-flow transient time also increases. We should note that low gas pressure should be used when testing adsorptive rocks due to the assumption of linear adsorption. In addition, gas desorption may induce matrix shrinkage which potentially increases the rock permeability [31, 32]. For these reasons, it is advisable to use non-adsorptive gases when testing rock permeability with CDT method.

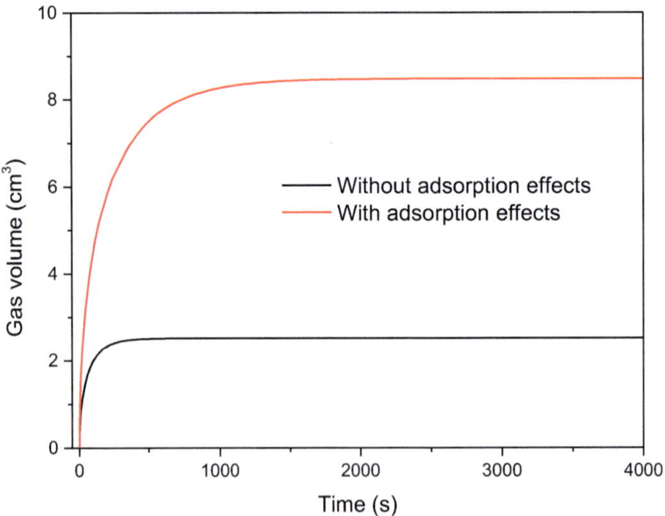

Fig. 6.14 CDT results with and without adsorption effects (pL = 2 MPa, Vm = 2.8 cm3/g)

6.4.6 Source of Error and Limitation

Several sources of error exist in the CDT method, such as system leakage, the measurement device, temperature fluctuation, errors from the determination of pore volume, and so on. Temperature fluctuation arises from the adiabatic changes in temperature caused by suddenly changing in gas drawdown pressure. To eliminate such effect, a small pressure difference between the saturation pressure and gas drawdown pressure should be applied. Since the CDT method records the gas flow data, a flowmeter with high resolution is needed in case of testing permeabilities of rocks with small pore volume and ultra-low permeabilities. Moreover, the proposed method of estimating the axial and transverse permeability is an inverse problem which may have non-unique solutions, therefore, reasonable initial values of axial and transverse permeability should be given to match the produced gas data (e.g. the permeability value parallel to bedding should be larger than the permeability value perpendicular to bedding).

6.5 Conclusions

An experimental method has been proposed for determining permeability anisotropy in this section. In order to test the validity of our method, we used a numerical simulator to generate gas production data. After matching the production data with our model, the axial and transverse permeability calculated using our model are very close to the true permeability values. We then carried canister degassing tests on two shale core samples from Longmaxi formation. The experimental results show that all the permeability values parallel to bedding are bigger than the permeability values perpendicular to bedding. We also compare the permeabilities results obtained from CDT methods with the results from PPD methods. The trend of axial permeability predicted with our method is closer to the PPD method, while our method can also yield the transverse permeability. At last, the effects of permeability anisotropy, porosity, initial gas pressure, rock dimensions, and adsorption on the profiles of gas production are investigated. The parametric study shows that the transverse and axial permeability affect the gas-flow transient time significantly, but have no effect on the cumulative gas volume, while the effects of initial gas pressure are on the contrary. For other parameters, such as porosity, rock dimensions, and desorption, can affect both the gas-flow transient time and the amount of gas produced.

References

1. Ettehadtavakkol A, Jamali A (2016) Measurement of shale matrix permeability and adsorption with canister desorption test. Transp Porous Med 114(1):149–167
2. Brace WF, Walsh JB, Frangos WT (1968) Permeability of granite under high pressure. J Geophys Res 73:2225–2236
3. Dicker A, Smits R (1988) A practical approach for determining permeability from laboratory pressure-pulse decay measurements. In: International meeting on petroleum engineering. Society of Petroleum Engineers. https://doi.org/10.2118/17578-MS
4. Jones SC (1997) A technique for faster pulse-decay permeability measurements in tight rocks. SPE Form Eval 12:19–26
5. Hannon MJ (2016) Alternative approaches for transient-flow laboratory-scale permeametry. Transp Porous Med 114(3):719–746
6. Kranz RL, Saltzman JS, Blacic JD (1990) Hydraulic diffusivity measurements on laboratory rock samples using an oscillating pore pressure method. Int J Rock Mech Min Sci Geomech Abstr 27(5):345–352
7. Wang Y, Knabe RJ (2011) Permeability characterization on tight gas samples using pore pressure oscillation method. Petrophysics 52(6):437–443
8. Potters M, Mansoori M, Bombois X, Jansen JD, Van dHPMJ (2016) Optimal input experiment design and parameter estimation in core-scale pressure oscillation experiments. J Hydrol 534:534–552
9. Cui X, Bustin AMM, Bustin RM (2009) Measurements of gas permeability and diffusivity of tight reservoir rocks: different approaches and their applications. Geofluids 9:208–223
10. Alfi M, Hosseini SA, Enriquez D, Zhang T (2019) A new technique for permeability calculation of core samples from unconventional gas reservoirs. Fuel 235:301–305
11. Luffel DL, Guidry FK (1992) New core analysismethods for measuring reservoir rock properties of Devonian Shale. J Petroleum Technol 44(11):1184–1190
12. Profice S, Lasseux D, Jannot Y, Jebara N, Hamon G (2012) Permeability, porosity and klinkenberg coefficient determination on crushed porous media. Petrophysics 53:430–438
13. Feng R, Harpalani S, Pandey R (2016) Laboratory measurement of stress-dependent coal permeability using pulse-decay technique and flow modeling with gas depletion. Fuel 177:76–86
14. Zhao YL, Zhang LY, Wang WJ, Tang JZ, Lin H, Wan W (2017) Transient pulse test and morphological analysis of single rock fractures. Int J Rock Mech Min Sci 91:139–154
15. Chakraborty N, Karpyn ZT, Liu S, Yoon H (2017) Permeability evolution of shale during spontaneous imbibition. J Nat Gas Sci Eng 38:590–596
16. Cao C, Li T, Shi J, Zhang L, Fu S, Wang B, Wang H (2016) A new approach for measuring the permeability of shale featuring adsorption and ultra-low permeability. J Nat Gas Sci Eng 30:548–556
17. Zhang W, Wang Q (2018) Permeability anisotropy and gas slippage of shales from the Sichuan Basin in South China. Int J Coal Geol 194:22–32
18. Chalmers GR, Ross DJ, Bustin RM (2012) Geological controls on matrix permeability of Devonian gas shales in the Horn River and Liard Basins, northeastern British Columbia, Canada. Int J Coal Geol 103:120–131
19. Pan Z, Ma Y, Connell LD, Down DI, Camilleri M (2015) Measuring anisotropic permeability using a cubic shale sample in a triaxial cell. J Nat Gas Sci Eng 26:336–344
20. Ma Y, Pan Z, Zhong N, Connell LD, Down DI, Lin W, Zhang Y (2016) Experimental study of anisotropic gas permeability and its relationship with fracture structure of Longmaxi Shales, Sichuan Basin, China. Fuel 180:106–115
21. Mokhtari M, Tutuncu AN (2015) Characterization of anisotropy in the permeability of organic-rich shales. J Pet Sci Eng 133:496–506
22. Yang Z, Dong M, Zhang S, Gong H, Li Y, Long F (2016) A method for determining transverse permeability of tight reservoir cores by radial pressure pulse decay measurement. J Geophys Res Solid Earth 121:7054–7070

23. Hosseini SA, Javadpour F, Michael GE (2015) Novel analytical core-sample analysis indicates higher gas content in shale-gas reservoirs. SPE J 20(06):1397–1408
24. Al-Hussainy R, Ramey HJ Jr, Crawford PB (1966) The flow of real gases through porous media. J Pet Technol 18(5):624–636
25. Li J, Liu D, Yao Y, Cai Y, Chen Y (2013) Evaluation and modeling of gas permeability changes in anthracite coals. Fuel 111:606–612
26. Moridis GJ, Pruess K (2014) User's manual of the TOUGH+ Core Code v1.5: a general-purpose simulator of non-isothermal flow and transport through porous and fractured media, Report LBNL-1871E, Lawrence Berkeley National Laboratory, Berkeley, CA
27. Feng RM, Pandy R (2017) Investigation of various pressure transient techniques on permeability measurement of unconventional gas reservoirs. Transp Porous Med 120:495–514
28. Wang CL, Pan L, Zhao Y, Zhang Y, Shen W (2019) Analysis of the pressure-pulse propagation in rock: a new approach to simultaneously determine permeability, porosity, and adsorption capacity. Rock Mech Rock Eng. https://doi.org/10.1007/s00603-019-01874-w
29. Zhao Y, Wang CL, Zhang YF, Liu Q (2019) Experimental study of adsorption effects on shale permeability. Nat Resour Res 28(4):1575–1586
30. Zhao Y, Wang C, Bi J (2019) Permeability model of fractured rock with consideration of elastic-plastic deformation. Energy Sci Eng:1–11. https://doi.org/10.1002/ese3.526
31. Lu Y, Ao X, Tang J, Jia Y, Zhang X, Chen Y (2016) Swelling of shale in supercritical carbon dioxide. J Nat Gas Sci Eng 30(4):268–275
32. Meng Y, Li Z (2017) Triaxial experiments on adsorption deformation and permeability of different sorbing gases in anthracite coal. J Nat Gas Sci Eng 46:59–70

MIX
Papier aus verantwortungsvollen Quellen
Paper from responsible sources
FSC® C105338

If you have any concerns about our products,
you can contact us on
ProductSafety@springernature.com

In case Publisher is established outside the EU,
the EU authorized representative is:
Springer Nature Customer Service Center GmbH
Europaplatz 3, 69115 Heidelberg, Germany

Printed by Libri Plureos GmbH
in Hamburg, Germany